Alien First Contact

Photographic PROOF of Alien Existence

Grant Irvin Miller

CONTENTS

PART II: PHOTOS AND CAPTURED LIGHT

PART III: THOUGHTS AND WORDS

Preface

The being I encountered is entirely real. I'm making this statement first and upfront, so there is no doubt in your mind that this may be some sort of trick. In later chapters, I will neither claim he's not real or alter my stand on this matter in any way. He was here, and I have the photo to prove it. There are also other images and information hidden in layers in the photograph behind what most can see on the surface. Once your eyes and mind become accustomed to looking not just at our normal interpretation of the standard dimensions of space/time, and matter/energy you may begin to see all which is there in the vast amount of information he's given us and you may also believe he's real. Of course, there will be many alternate theories, skeptics and debunkers who will offer explanations ranging from children running around in Halloween costumes to copious amounts of swamp gas. The swamp gas theory is probably the most plausible and closest to the truth, and I admit was likely created by my brother's baked bean casserole prepared for our backyard BBQ a few days before the picture was taken. I love his baked beans, but they always have the same effect on everyone. If you must believe that I took a photo of gas, so be it, but if you're a fairly open minded, intelligent person none of the alternatives will strike you as true and so I will attempt to prove to you just how real these beings are.

Staring into the sun is pointless though, at least for humans. You can do it for a second or two, but sooner or later your eyes begin to hurt and loose focus and you have to turn away. Once you do, you will naturally avoid ever looking directly into it again. The best way to deal with this is to look only at the places which the sun illuminates catching a glimpse of the intense brightness out of the corner of your eye. That way you can slowly get a sense of what the sun looks like without all the pain. In time, you can put the glimpses together into a pretty good idea of what you would see if you stared unwaveringly into the full ferocity of the star.

You'll see I use a fair amount of humor as well as drawing many parallels while telling the story. It seems to be the best way to approach many of the ideas here for much the same reason as I outlined above. I found it far easier to give an understandable description of many events without making them too strange or frightening. Frightening, I've found, is usually an illusion. An illusion to be overcome.

This all works pretty well with aliens in general and especially with this photo which seems to have a powerful life of its own. It is merely a photograph but quite often affects people in strange ways. Simply showing someone the photo and bluntly explaining what's in it seems to produce odd reactions and many times will cause their mind to dismiss it completely. Most people just do not want to believe it is real. So why would I write this book if almost no one will believe it or ten years from now even remember the book existed? Because this story needs to be told and the photo studied. Because once again, it is real, and it won't go away, and so if just one person believes, then in time, more will follow.

Take a huge breath.... we will go very deep. The currents crisscross in all directions. I'll try to answer as many questions as I can. Questions like; How did I come across these aliens? Why can't we see them clearly? How do they communicate and why can't we understand them? Why do people forget or deny they see them? How do they travel? How long have they been here? Do they have sex?

And by the way, I needed to get this information out quickly and so settled on self publishing as the fastest and most efficient way to do so. Therefore I did not have access to a traditional publisher's army of ghost writers, copy editors and proof readers to correct everything from missing commas and dangling prepositions to reordering story flow issues and cover design, or the two to three years it might take to get the book to readers. I believe I have caught 98 plus percent of all issues. The information itself is much more important than the way in which it is presented. I wrote this book just the way the information poured out of me, and boy did it pour! I did the best I could on short notice so please excuse minor grammatical errors and such.

Part I deals mostly with background science and experiences. You will need this information as a base to understand Parts II and III and believe me it's worth the effort. If you are looking for answers, descriptions of aliens in the photos, and much, much more, hang on for the ride of your life. This a true story!

There are more books to follow. It all just keeps unfolding.

And it's quite a journey you're are about to embark upon.

Stay tuned and stick with me.... Here we go!   and Oh Yes.......time to put your sunglasses on.

Part I

The Visible World

Chapter 1

A Day In The Life Of A Dog

It was very much a typical day in the summer of the year 2000, and I expected nothing out of the ordinary to happen. I arose and found my way to the refrigerator to get myself a glass of milk, and I looked in a sleepy fog at the food there attempting to decide what I would eat for breakfast. I shuffled the contents and when nothing leaped out at me so to speak, and being rather disappointed that I had found no eggs, bacon and toast already cooked and waiting for me, I closed the door and turned to the cupboard to get the cereal box. My dog Greggor was awake and alongside me now, and probably felt he would miss out on something important or delicious if he stayed curled up in a ball. He happily padded up to my side and sat waiting for me to choose my breakfast. Greggor was a wonderful friend, curly coated, and very bright. He was usually stuck to my hip, and would wake from a nap to follow me to the next room if I moved unexpectedly. He was also a complete pain in the ass to keep clean. He needed constant care and grooming, and if I neglected him for even a day he would demand my attention until I properly brushed him, removing all the old shedding hair and burrs and ensuring his feet were clean and free of foxtails. Adopted from the local animal shelter, he had been quite sick when I found him, and it took me considerable time to convince the workers there to let me take him home, but only if I would stop at the veterinarian's office and have him attend immediately to the dog's needs. When I first saw him at the shelter, he was horribly skinny, coughing and listless, but he managed to stand long enough to draw my attention and I immediately formed a sort of bond with him. His hair matted to the point that it looked like muddy dreadlocks completely covering his body; he was an utter mess and was going to take a huge amount of work to salvage, but his personality or some spark or something shown in his eyes so strongly that I could not go without him. His personality turned out to be just what I had sensed, strong minded, smart, courageous when needed but generally cool and calm in manner. He greeted all humans as if he had known them as family since birth and approached most all animals the same way. I remember one evening when one of the family of racoons who lived nearby and scavenged for cat food left unattended overnight found it's way into my yard. Greggor confronted it on the lawn halfway from the back fence to the house, and the raccoon froze and squared off. Greggor walked up within two feet of him and neither barked nor growled or teased to chase but instead sat down and watched respectfully. Then he looked up at me as if to say, "Look what I've found!" and turned back towards his guest. The raccoon, after regaining it's nerve, slowly padded off back towards the fence and disappeared underneath.

Breakfast! Yes, and I was hungry! I knew Greggor was there but paid little attention other than a friendly pat on the side of his neck. As I peered into the refrigerator though, he turned his head toward the back yard and let out a small woof. He paused, cocked his head as if a bit confused and then turned back towards the allure of food. The morning sun was strong and bright. The air was already warming, and it promised to be every bit the hot day forecast by the weather service. I could say I expected nothing out of the ordinary to happen, but had I taken more notice of the events leading up to this day, I may have had a different expectation. The summer weather, to me had seemed kind of strange, but I had been seeing stories on global warming and climate change on the nightly news programs and so I quite expected a few minor changes to be apparent, but nothing much had happened as far as extremely high heat or blisteringly dry weather. California summers can be long and hot, but except for a few degrees variance from previous years there was nothing unusual going on. It had not rained cats and dogs nor had frogs or flaming cows fallen from the sky. At least not that I had noticed. But sometimes I thought the light looked a bit odd.

At this point in my life, I really did not have much belief in UFO's, alien beings, crop circles or paranormal anything. As far as I knew crop circles were a new breakfast food. I watched the latest science fiction movie hits like most everyone else, but the reality of aliens and UFOs was not something I bought into. Now I did later get into watching The X-files on TV and loved to watch Mulder and Scully battling all sorts of X-stuff things, but that was later, reruns. At the time I encountered the "being", I was working so many hours that I did not even have time to catch many TV programs at all let alone my favorites. Later on after my encounter I became obsessed with the cosmos, cutting edge physics theory and discoveries, ancient alien astronauts and paranormal research, plus good food and travel. So getting back to my story, I really had not consciously noticed anything startlingly unusual at the time, but thinking back I am quite sure that there were many unusual things going on, which were probably due to the aliens interacting with the environment in preparation for contact. And what does that mean? I can only relate the strange and unusual which I recall. But first of all, I can not remember some nights during that period. I have scoured my mind and just can not recall what I was doing on certain nights! I have blank spots, which stand out in my mind as unusual. I remember work and getting home and eating and then nothing, nothing at all until I woke up the next morning. The interesting thing is that I remember often waking up the next day in one freaking great, fantastic mood and wondering to myself whether I had a good sex dream or something. Maybe exercise from the day before relaxed me so I slept unusually well? But no, I was so busy at work that my physical health was already calling to me to get back on a regular exercise program. I subconsciously understood that something was different, and probably knew what it was but could not bring it into focus.

One day while I was reading a novel I recall hearing sounds by the side of the house, strange and different sounds, muffled thuds, a whirling noise and voices. The voices, or so I call them, sounded like human voices, but I could not make out any words at all. The floating syllables and isolated consonants wafted through the air, and I would be certain they were in sentences but could not quite grasp enough to put the whole collage together as if it were all just beyond my reach. It was all odd enough that I ran outside and around the side of the house. The plants were thick, but I plowed through with determination. But there was nothing there, and no voices to be heard either. The neighbors were not home that day. I had checked immediately. Racoons perhaps? Fairies? Possibly, I suppose.

There was also an incident which I remember that happened one night when I was in bed. As I lay there falling asleep, I woke to the sound of footsteps which seemed to be coming from the rooftop. I waited and listened and then again, there they were one after another after another, step after step after step, and then they just stopped. Right above my head, they just stopped! I remember being intensely interested, and a bit scared, wondering who would be on my roof! I was so convinced that I wanted to go out and get a ladder and immediately climb up and rush towards where the steps came from to see who it was, but all that would take way too much time and effort. And so I listened for a moment more and eventually fell asleep. Many times I awoke to feel that someone or something was watching me through my window. I would turn my head and listen. I would try to focus my eyes in the darkness, not daring to open them completely and watch intensely out the open window to see any small movement of creature, tree or bush. I sometimes watched for fifteen or twenty minutes just to try to catch whoever was there, but I never did. There was never anyone there, and I eventually forgot all about the odd events. For a while at least.

I was working out of town and after a long commute home, and some really poor fast food, I arrived at home with indigestion, and approached the front door only to find it unlocked and open. I remember muttering to myself something like, "What the hell is going on here?" But upon entering I found nothing missing and only a few small items were possibly even out of place. I went to the refrigerator and to my amazement it's door was also ajar. I looked inside and noticed not much except that the milk was open. Milk. Open. Alright. My only thought was that someone had been in my house! Either that or I'd been burglarized by a cat! Was I losing my mind? It didn't look like any milk had been consumed. It was almost full the night before and still was. I woke up late that morning and jumped into the shower, threw on some clothes, and hurried off to work. I had not even entered the kitchen. Or did I actually go in there and later simply fail to remember? Did I leave the door open, the refrigerator open, and the milk open, and then just drive away? I retraced my steps and was positive I hadn't. I'll probably never know for sure but a few more days after that the really weird stuff started happening.

One nice warm sunny day, I arrived at home, unlocked the front door and went in. After a few minutes I walked in towards the rear of the house. My bedroom is there, and I could see from well in front of the doorway, that, for some reason, it was brightly lit inside. Moving into the doorway itself I could see that the light from, the sun I assumed, was streaming in through my window. Not so strange you might say except that the blinds were closed! The light beams were so strong that they literally pierced through the fabric of the blinds. They illuminated the room to such an extent that I could not focus due to the glittering and shifting light flooding in. I stood in the doorway, and I stared. I felt a weird sense of awe come over me. I wanted to take a deep breath but for some reason was holding it and could not let it in or out. I had never before seen sunlight like this in my house, or for that matter had never seen anything like it. And the light was a weird yellowish white. No combination of the sun setting, sun spots or reflections off other objects could account for this display. Bewildered, and standing there for some time with my mouth gaping open, I watched as it faded away. The strange light just seemed to melt into the ordinary afternoon glow and was gone. Everything was silent at that moment.

I think back on those few minutes I spent transfixed as I stared into the room. I really don't know how long I was standing there. Have you ever suddenly seen something, so unusual, that it strips you of all will to move, breath or interact? It commands you to stand motionless, inert, without thought or incentive to be anything or anywhere, except where you are at that moment? Frozen. I try to describe the dimensions of this "awe" but always fall short, not being able to relate it accurately or completely. I suppose I must again sum it up as everything was silent at that moment, including me.

I regained my presence in the room. I was aware that I was again breathing.

I then looked down at the lower part of the wall as something caught my eye. I blinked and looked again but couldn't see anything strange or odd, nothing moving. Why was I looking at a blank section of a white wall? A rattling noise then broke through the silence, and I raised my gaze. The window had rattled for some reason with a slight buzzing, a vibration reminiscent of high speed dentist drills or machinery running, but slightly lower in pitch. It stopped just as quickly as it started. I returned to the wall and way down, a few inches above the base a wisp of plaster dust came spraying out into the room forming a small dust plume. I say, spraying out because it was almost as if something had pressurized the inside of the wall, and was now drilling it's way through! This weird pressurized wall was springing leaks! I panicked. This was fairly new plasterboard. There were no cracks or seams, and there were no other holes or marks of any kind on it. Another wisp of dust spewed out as I watched. It was an exterior wall, so I hurriedly ran outside to see who or what was there and how this was happening. When I reached a spot that I thought was opposite the site on the interior wall I suddenly and stupidly realized that the exterior surface was stucco! Half inch thick stucco mortar mounted on chicken wire, over wood planks. There was no way anything was drilling into my house! It takes a special mortar drill bit just to make a mark in the surface. No animal could have bored in, and I could not find a single entry point. It was solid, smooth and untouched. I have never had a problem with insects or vermin ever attempting to get in. Not through the front or rear doors, and certainly, not through the freaking wall! There were no windows nearby, and the wall itself bottomed out into a solid concrete foundation! I panicked again. I did not know what to do. Visions of crazed mutant monsters from the movies began slowly breaking into my subconscious. Hideous things were alive just inside my walls, just beyond where I sleep! At this point, I was not thinking clearly and ran to the storage shed in the backyard. It was locked, but I realized I had the keychain in my pocket. I fumbled to find the right key, and pushing it into the lock, I opened the door and grabbed a caulking gun. It still had a half used tube of bathroom sealer in it from the week before when I had recaulked the tub. I suppose having the gun made me feel safer. I ran back into the house and into the kitchen where I opened the drawers one by one. Rummaging through the utensils I grabbed an antique ice pick. Yes, an ice pick! Now I felt even safer because I could really defend myself! I ran back to my room where I once again froze at the door. For seconds, I watched. This time though, I was alert. I stared at the wall where I had seen the plaster dust plume. To my relief the hole was much the same size, so I slowly and quietly approached. The light had faded, and the vibrations had stopped, so I stared again for a while to see if anything moved. It was about the size of a small nail hole and material had sprayed out of it leaving small bits around the opening. I slowly knelt down, and on my hands and knees, crawled closer. All was strangely quiet once again. I could hear only my own breathing, fast and hard as one breath replaced the other. My bare knees stuck out beneath my shorts, and they sunk deeply into the rug, landing on a small stone hidden there which bit into my kneecap. I dug out the stone from the carpet as I stared at that hole. I never looked away. I did not want anything to climb out or squirt out, and I desperately wanted to seal it up as quickly as I could. I crawled one step closer. Then I froze. Suddenly, behind me, I felt a drop of warm liquid goo hit the top of my bare leg. My mind went back into panic mode as I imagined some hungry monstrous creature or alien about to sink its teeth into me! It was just behind me. I was sure I was a goner. I couldn't move! I couldn't turn to look! More goo landed on me! The liquid slowly dripped down around the curve of my leg and I imagined my own blood following shortly thereafter. So many things pass through your mind at a time like this, and I wondered what it would be like if by some chance I survived the attack but lost my leg, crawling away, fighting the vicious creature off with the ice pick and caulking gun. My leg! My leg! What if it got to my throat? The light changed slightly! It was time! My time to die! My time to be eaten by the thing reaching out at me, standing right behind me! My frozen panic suddenly broke, and I was able to move once again.

Slowly I turned my head sideways and jumped up, shocked to find Greggor standing behind me, drooling! He'd been drinking water moments before. I never even heard his approach, which is so unlike me having known the dog and his noises and mannerisms for some time. I had dropped the ice pick and caulking gun when I jumped, and I looked around to find where they had landed. I did not want to get any closer to the buzzing drilling in the wall, but the ice pick landed and rolled in just below the hole! I once again dropped to my hands and knees, and seeing that Greggor was still with me I began to crawl forward. But Greggor was not moving. In fact, he was not moving at all. I slowly turned my head, not wanting to lose sight of the hole, and found that he was frozen as if paralyzed in time, staring at the same place I had been. I grabbed the ice pick and I lurched at the wall slamming the pick deep into the orifice. It penetrated easily but then found nothing but empty space inside. I pulled it out and slid it back in just in case there was something alive in there. I jumped back away and waited as Greggor stood frozen, not moving, not blinking, and not breathing that I could tell. I couldn't look at Greggor now. I once more stabbed the pick in and twisted it around to open the hole, but then realizing that nothing was in there, I stopped. I waited quietly for a moment more. No sound. No movement. Nothing ran out of the hole. I got up, and went into the next room to get a flashlight. I remember walking past Greggor and getting the flashlight, then passing back next to him again. He didn't move a muscle. He looked like some strange stone statue to me. It all seemed as if it were a dream. Was I dreaming? Would I wake any moment to find myself surprised at being in bed? I fell to the floor, and shining the flashlight into the void, looked deeply inside but found only empty space. I grabbed the caulking gun and immediately pushed the tip to the hole, squeezed the trigger and filled it permanently! Greggor moved in next to me, and I realized that he was there and patted his neck in relief. We had both survived.

Chapter 2

If They Were Easy To See,
We Would Have Already Found Them Long Ago

In my life, I have often found many things that seem to possess the ability to become invisible. As a child, I could hear the call of a robin nearby but as much as I looked I could not see the bird in the trees. He would call and call, probably alerting other birds to the presence of a cat in the swaying grass beneath the tree, but as I strained and moved back and forth I was only able to discern that the bird was so well obscured from my vision as, to actually be, invisible. Now most likely I was not looking in the right place, on the correct branch or between the right leaves. I was just an innocent little kid who thought that I should be able to see every little thing which was in my yard and if I could not see it then it was truly invisible or did not exist at all.

I remember fishing in a small stream about a mile from my house. I went to the same spot quite often because the other kids who fished claimed that they had seen the shape of a large trout in the shadows, bigger than any other ever seen here, slipping into a hole beneath a tall tree in the bend of the stream. I waited hours to catch a glimpse of the fish. I was sometimes there at dawn, noon or sunset always watching and hoping to see the huge creature. One day, after all but giving up hope, I sat down laying my pole by my side, just above a smooth spot on the bank, just above the hole. The sun streaming through the trees cast eery shadows across the ground and also partially illuminated the water there. I looked and blinked, and there I saw a shape, a large shape beneath the running water! It seemed to move six or eight inches toward an underwater root but the changing light obscured it and seemed to turn it into a rock. Or did it? Was the fish really there, or was it my imagination playing tricks on me? I quickly crawled to the edge and put my face nearly to the water and peered intensely into the deep. It had to be real! But there was nothing there that looked like a fish, no movement, no scales, fins or, and then, *Splash*! The fish, sitting half hidden but camouflaged naturally against the rocks near the opening of the hole, found itself too exposed. It turned and whipped it's tail fin upward, propelling it down to the depths, breaking the surface as it did, and scaring the living shit out of me. As I knelt there with a shocked expression on my face, I realized that I never really did see the fish. I imagined or just knew that what I thought happened was what happened. It agreed with the facts of my experience. Was it there in the shadows? Had I been looking right at it all along but could not see it? I don't know.

Many things in life seem to be almost invisible at times. This, for the most part, means that we are unable to see them, or experience them right now, from the vantage point that we have. It does not mean that we will never be able to see them or that they do not exist. Not that they are actually invisible. But you could easily refer to them as invisible, and no one would know the difference. Strange as it may seem, there are invisible objects, and invisible experiences. And they are being used by others, to enable them to mix with us, examine us and our way of life, and do it so well that almost no one would believe they were here at all. And they may have been doing it for a very long time. They are completely inconspicuous, or so it would seem to us as we blunder along through the bizarre world of the invisible where we see only what we feel is important at the moment, and ignore that which seems of no consequence to us.

Invisible things could be categorized as being of three different types: Mostly physical camouflage, mostly psychological camouflage and truly invisible. And why all of a sudden did camouflage get mixed up with invisibility. Because most things we think of as invisible, are actually just extremely well camouflaged. The physical universe also has lots of nooks and crevices which we can not easily see into. So hidden there, are things invisible? Depends upon your definition of invisible! Some things are too small to see with the naked eye. But with an advanced microscope, their entire world comes plainly into view.

Definitions, yes, definitions.

Chapter 3

The Language of Camouflage

I wrote this section with a fair bit of awkwardness. Although hundreds of years of study by military experts, psychologists, magicians, writers and con artists has given us a wealth of information on the subject of both camouflage and invisibility, I have yet to see a comprehensive guide to their combined use and many aspects of which are actively kept secret by the same groups who have done the research experiments and studies. To some extent I will be inventing and theorizing as I go along, but I will try not to stray too far from accepted science. So first let's discuss some terminology because just using a word rarely defines it the same way in different peoples minds. All this relates to camouflage and invisibility and how our minds work in some way. It won't take long and hopefully it will be fun.

Mostly -- Means, almost entirely but not completely or exclusively, something like that. Example: The tree is mostly green. There are also some brown leaves and bark which we shall not call attention to right now.

OK. I admit that was just a warm up! Not what you expected! Seems ridiculous and pretty fuzzy, right? But just keep going. You have to think outside the box to get all this. So take another deep breath, and since you've already made the decision to keep reading, please stop objecting and proceed to the next line.

Mostly Physical -- Means it applies to the physical makeup of an object or being but is not exclusive to those physical properties. Effects may have a psychological counterpart, as well. Example: The deer's legs look like sticks in the grass. It's brown body seems hidden against the trunk of the fallen tree. This mostly physical effect blends it's body coloration into the surrounding background making it nearly invisible.

Mostly Psychological -- Means that it applies to the psychological makeup of the situation or subject matter and generally is realized that it's all in one's mind, but may also have a physical counterpart or trigger. Example: The mentalist magician pointed and gave you a subtle, mostly psychological suggestion to examine the birds in the grass closely and thus drew your attention from the deer whose legs and body were hidden against the trunk of the fallen tree.

Purely Invisible -- You can not see it or detect it in any way. Example:

Camouflage -- The disguise of items. Camouflage can be said to utilize physical things like paint, colored body parts, projections of the surrounding landscape, dead leaves, skin texture, etc, to make an object or animal less visible. Example: The wavy camouflage pattern painted on the truck makes seeing it as a vehicle difficult. It looks like a giant jigsaw puzzle.

Camouflage may not fool you or anyone else, but if it's really good, it just might. It depends how your mind works or perceives reality, and what your level of expertise might be. If you are fooled, it's your mind that has been fooled. The object is really there. Right? But your brain does not at that instant recognize that there's anything unusual there beyond the basic background, brush, trees, animals and insects. Specially trained military personnel, hunters or just well informed individuals can much more quickly see through the effect or not be fooled by it at all and so the matter of detecting camouflaged objects is all but a matter of being trained to do so and having an open mind.

The "fuzzy" use of these terms is again because invisibility and camouflage are by definition (in my mind at least), mostly pretty fuzzy by nature. They're nebulous and not so easy to define so that everyone has the same idea. And because none of them adequately applies well to my photograph, we need to all be sliding across the same page as much as possible.

Invisibility is also subject to debate. Can something actually be invisible? Cameras routinely capture objects which are unseen to the human eye. Visible light, or the light that the human eye can perceive is only a small portion of the electromagnetic spectrum. Attempting to stay clear of a lot of heavy science, the human eye has a range of or responds to wavelengths of about 400 to 700 nm. A "nm" is a nanometer or an extremely small measure of distance. Many infrared and thermal cameras have a range of up to 14000 nm. Infrared light is invisible to humans, and so these devices can capture much more information than our naked eye. Infrared cameras and sights can capture images in what would look like total darkness to us because they either see the heat signature of objects or animals within their range, or intensify available light. So if you can not see an animal with your naked eye, is it invisible? Maybe. Maybe not, but it's English terminology and it's fuzzy.

The relative velocity of the subject is yet another facet of the entire issue. The faster an object travels in relation to it's distance from you, the less likely you are to see it. Cameras once again can capture things which are happening so fast that they do not even draw our attention let alone allow us to see and comprehend them. Many photos have been taken where the operator sees nothing in the viewfinder only to reveal an insect, bird, projectile or UFO upon later inspection of the captured image! Could you see a bullet speeding by fifteen feet in front of your eyes? No, But high speed photography can show us that detail. We have all seen a humming bird's wings become frozen in time. A bullet standing still in mid air. The whirling propellers of an airplane made to stand still in space, we imagine they spin their way around the central hub. A drop of water hitting a pool. The splash. The ripples.

Physical things can be made to become mostly invisible in many ways. Stealth fighters have been engineered to be almost radar invisible through the use of radar dispersing body elements and radar absorbing alloys. The radar signature of a stealth fighter is closer to the size of a hummingbird or fish as opposed to the radar signature of a plane, which is comparatively like the size of a whale. The plane can rarely be seen from the ground with the naked eye because it flies very fast and high. Odd facets on the planes body make identifying it by sight extremely difficult. This mostly physical camouflage works from almost any angle. The jet looks like a flying puzzle piece and appears to change shape as it turns. Camouflage has been used for many decades to make aircraft and ships, as well as large weapons, appear less visible. Used in times of war, this technique has proven very valuable in fooling the enemy into thinking these objects were not actually there or were something else altogether.

Chapter 4

Invisibility & Camouflage

I once read a story, The Invisible Man, by H.G. Wells, originally published back in the year 1897. The main character of the story is a scientist named Griffin who becomes invisible during an experiment and then finds that he can not reverse the effect. Griffin's theory is that if a person's body is made to neither absorb nor reflect light, then he will be invisible! His experiments are successful, but his life as an invisible man proves to be much more difficult than he anticipated. His temper becomes a problem. He encounters different people whom he intimidates or accosts while they can not see him. He sometimes throws objects in an attempt to frighten those around him away and steals to survive. It's just a story I know, but maybe there's more to it than just literary imagination. What if it could be true? What if there were a way to make yourself absolutely undetectable to other humans, and allow you to go, or do whatever you wished, whenever you wanted? If it were actually true, where would you go? What would you see? What would you do?

Well, It's only a story for us humans perhaps, but for others, possibly not.

As we have all learned in grade school, some chameleons can sense their surroundings and adapt their coloration to objects nearby, or to objects on which they sit. Their skins are covered in specialized cells. The cells, which are set in layers under a transparent skin, can be directed to change their coloration, and there is a wide range of colors which they can use. The changes are most likely for signaling other chameleons, but they can also be used for the purpose of camouflage. A chameleon in camo mode can be quite difficult to see. Their skin coloration can blend into the foliage to the point where an observer simply passes by them without notice. But as incredible as the chameleon's performance is, there are creatures on our planet whose skills of camouflage are utterly amazing! The octopus is one of these. Their ability to change colors is very well documented. They can not only change the color of their skin, but they can also use the muscles beneath to change the textures of those colors created as well as the opacity and reflectiveness. The octopus is among the most intelligent of all invertebrates and uses tools. They can be trained to mimic shapes and patterns and have been observed to change them as if at play! Some species of Octopus combine their highly flexible body with color and pattern changes to accurately simulate the appearance of coral and other sea creatures.

In nature, they literally embody the phrase, now you see it, now you don't!

The walking stick, also known as the "ghost bug" can sit on a small branch just in front of your eyes and not be seen. You can look right at it, and it's just not there! It's just a piece of the bark or a twig until it moves and then everything changes in your mind as you realize you had no idea what was right in front of you. In other words, you realize that you had no idea what you were looking at when you were looking right at it.

The leaf mimic frog looks so much like a dead leaf as to make any frog hunter crazy.

The praying mantis adapts its coloration to many different plants and environments and is among the masters of camouflage.

Hunters and woodsman have long known that camouflage clothing allows them to be in the woods and be almost visually absent as far as animals are concerned. Some of the newest camo gear is absolutely amazing! Nothing short of spooky. If I were in the woods and not aware of a man in camo in the same area, and he was doing something like throwing stones or making noises to make me nervous, I would seriously consider that I was being haunted. Some of the newest hunter's clothing is mind boggling. One new concept of the best companies creating camouflage designs involves asking for photos of the woodlands where the men will be hunting. They can then take those photos and create a collage of colored images which can be printed upon clothing, etc. Consider the depth of the design and the complete coverage of a man camouflaged with this technology. Gloves cover his hands and make them look like leaves, grass or brush. His weapon resembles a branch or twigs, and a mask covers his face. He is virtually invisible to his prey.

These are all examples of mostly physical camouflage or invisibility, the use of physical attributes to alter or confuse the observers sensory data. We do not see them as they actually are or see that they are there at all. The effect is that all one is able to comprehend in his mind are colors, leaves, grass, twigs, bumps and clumps of stuff, flashes of light and nothing that one's mind needs to pay the least little bit of attention to. Everything in your world is exactly as you expect it to be, and the camouflaged animal is just not there or part of it. If they're friendly, your space is safe, perfect and stable. The whole point of this, is to remind you that many creatures here on earth, can effectively disguise themselves to evade, elude or confuse us, and so could intelligent, sentient beings.

Scout Sniper in Camouflage

Pfc. Travis English, a scout observer from Echo Battery, 2nd Battalion, 12th Marine Regiment, practices observation techniques using cover and concealment techniques and a homemade ghillie suit.--   Photo by Staff Sgt. Marc Ayalin – Photo Courtesy of DVIDS

Chapter 5

Human Chameleons

The techniques of expert snipers and scouts is astounding. When dressed in camouflage, snipers are at best difficult to recognize as human. In heavy foliage the "ghillie suits" they wear are customizable to their surroundings. These airy suits are often made of burlap or twine and can be dressed out to more closely resemble the foliage they will be moving through. They move very slowly and never assume their gear will continue to match their surroundings, so they may stop and carefully choose bits of brush and grasses around them. replacing pieces attached to their suit as they crawl! From a distance, the enemy would be completely unaware of it. The most amazing part of this is that as they move there is no way that you could detect them without first knowing their position. I am not certain I would find them, even then. They can move so slowly as to mimic the breeze moving the grasses. The trek across the expanse to their final position could take hours. I am truly amazed at the level of expertise and stealth they demonstrate. So can we be fooled into thinking that something near us is not there? You bet! But if we do realize it's there, there may also be several ways of making us forget about or dismiss what we have seen or heard.

Marine in ghillie suit

A Marine tries on his ghillie suit after harvesting grass to vegetating it during stalking training conducted at a remote training area near an inactive shipyard at Joint Base Pearl Harbor - Hickam. The snipers are students of Scout Sniper School, School of Infantry — West, Detachment Hawaii
Photo by Lance Cpl. Tyler Main Photo Courtesy of DVIDS

Chapter 6

The Art of Psychological Camouflage

As every magician knows, the hand is quicker than the eye. The flip of a card can be done so fast that the human eye can not follow it. Now you see it, now you don't. Relative speed equals invisibility. A good magician will at the same time also misdirect your attention to another point to ensure the trick is truly invisible. He will possibly have his pretty assistant pose strikingly just before the card flip drawing your attention to her instead of the card in his hand. But suppose he doesn't pull off that misdirection well at all? Perhaps his assistant's timing is off, or her skimpy costume is at the dry cleaners that day. What else is there to ensure you do not catch him red handed. A flash of light, a smokey room, a dazzling stage backdrop? What else indeed?

Let's delve into psychology for a little bit. Let's look at, for example, a dream, an amnesia..... perhaps a fugue.

The fugue. According to my memory of psychology, a fugue is

a condition during which you are apparently conscious of your actions, but you have no recollection of them afterwords. This condition usually results from severe mental stress, and memory loss may persist from as long as many months to years. It could also be seen as a dreamlike altered state of consciousness lasting for long periods of time during which you lose the memory of previous events. There's also something called dissociative amnesia. The central cause of dissociative amnesia is stress associated with traumatic experiences that you have either survived or witnessed. So what we have here in general is the notion that when in highly stressful situations, I mean under unusual stress, such as the stress of seeing an alien being or UFO, many people may afterwords forget what they've seen and may never consciously want to remember it again.

Many people who see my photo seem to "blank out" after a while. They either can not see the subject at all, they forget shortly after seeing it, or don't remember a thing they've seen several days or weeks later. Why? Perhaps a sort of fugue state? A dissociative amnesia? Whenever I show someone the photograph, I start by attempting to sound them out as to their religious, spiritual and paranormal beliefs and also to ascertain their level of scientific thinking and openness. Then I try to put some context to the photo and finally I attempt to calmly, quietly and very nicely show them what's there. People then react in different ways and tend to fall into a few loose groups. Three of these groups are worth mentioning now.

This first group cannot see the being at all and sometimes get quite disturbed or angry when I attempt to point it out to them. Once released from the conversation and proximity to the photo they then relax and are calm again but become agitated once more if even reminded of the subject matter hours later. I believe that their subconscious mind clearly sees the alien being but for some reason rejects it immediately and then restricts access to the visual memory information. Consciously they are left in the dark and insist they are correct that there is nothing more than my dog in the photo. Why would they become agitated? Why would people avoid discussion of the photo? Possibly because we have over time developed a safety valve in our minds. One which says, that if I can't see this thing and it doesn't outright hurt me, I'm better off if I just don't consciously realize it exists.

Interestingly enough there is also a musical style called the fugue.

The fugue in the musical sense is a type of composition in which a theme or themes are repeated over and over again, each one laid on top of the other. How interesting that a word associated with memory would also be used to describe a musical form with repetitive overlapping verses. The word itself was most likely taken from the Italian word fuga which means flight. One musical theme overlapping the first and a third overlapping the second and on and on. The effect can be beautiful to listen to, or completely nerve racking depending upon how the piece was written or performed. Memory has some elements in common with this form of music. To ensure a memory is retained we repeat or overwrite that event again and again, reinforcing it if you will. Practicing scales on a musical instrument such as a piano is a good example. The more you play the scales, the easier they are to remember. If one sees something which is disturbing, the subconscious mind may also attempt to "hide" that event by overwriting it with a layer of memories from a point when it did not exist. It could be that people not wishing to "see" the being immediately start covering up that vision with a memory of the photo from just before they saw what was in it. Each layer overwriting the previous layer, and when interrupted in the process of doing so, they then become agitated and angry. Have you ever heard a repetitive "singing" in your sleep. We will get back to this at some point.

A second group of people sees the beings in the photo and holds onto the information for a few hours or days but shortly after that they reject the fact that it is real and attempt to forget all about it. They seem to experience an amnesia which gradually erases or overwrites that memory in their minds.

The third group is the smallest. They see the being immediately, tend to grasp what's going on and have no trouble with believing it's real. They also retain the memory. Hopefully you're one of them. If not, there is a place you can go where you can face these unimaginable monsters and strange events, where life and death merge into a kaleidoscope of multiple colors, feelings and experiences, and you can emerge unscathed, unharmed, and remember not a thing moments later. Dreams.

Sleep, unconsciousness. Do we ever really remember falling asleep? I mean, the actual moment when we lose consciousness? That instant between making yourself comfortable, closing your eyes, clearing your mind and then, nothingness. Gone. Asleep. No? Most people do not. Some people attempting to stay mentally cognizant up until that event horizon say they experience a strange effect at the last moment. They claim they see what appears to be a vortex made of swirling dim light trails in shades of white which appear just before they start to spin into it, and slipping away towards it's center, begin to sleep. Could this be why we call it falling asleep? Falling into a deeply relaxed state, we seem to enter another place altogether whether in our minds or possibly in reality. And could aliens, if normally camouflaged to us humans, find it easier to communicate with us when we are in a very relaxed, possibly suggestible and vulnerable state? In the dream state, we are also essentially paralyzed. Could they take control of us during this time to give us some alien direction? Maybe they only want to be able to examine us when we are much less of a danger to them. Is it possible that this is going on? That's the question. Isn't it? Many alien abduction accounts speak of night visitors who appear from nowhere. Many abductees can not recall details. Possibly the sight of aliens causes a fear reaction that makes it difficult to remember. Coupled with sleep paralysis, this could provide the aliens with the perfect place to communicate to and study people without rejection and without us remembering. Perfect psychological camouflage.

Chapter 7

Sleep and Dream Time

The human brain contains upwards of a hundred billion nerve cells and chemicals called neurotransmitters which control whether we are asleep or awake. When they are activated, we become drowsy and fall asleep. Upon awakening we promptly forget the entire experience including our dreams. Sleep induced amnesia is a common regular occurrence and could also be described as, "I really don't remember anything of what happens while I am sleeping once I wake up!" Then again, most of us do not remember our dreams, or very rarely since we all dream every single night. So we are naturally in an ongoing state of amnesia during the hours we are awake. We forget everything that happened in our dreams the night before, every morning. Once more, we all suffer from this memory loss, every day of our lives. There are many theories attempting to explain this, but no one has the complete answer.

And what might we not be remembering? For at least eight hours per night we sleep, we dream and are essentially paralyzed and almost never remember any of it. This would mean amnesia is a normal brain function which evolved for some strange reason. We are paralyzed, so we don't move, and experience amnesia, so we don't remember. We are obviously quite vulnerable and could easily suffer or encounter stressful situations. With small adjustments to our brain chemistry, a minor fugue effect might be capable of camouflaging everything that happened to us no matter how bizarre or alien. An alien interaction perhaps? And so I ask, "Could this be happening somewhere right now as part of some alien plan?" Maybe not to everyone every night, but to some, intermittently or regularly. Bizarre you say? Abductees often resort to hypnosis in an attempt to remember the events of the night of their abduction. Could it be that aliens engineered us to sleep for this reason thousands of years ago? Regardless of the answer, we don't know what they're up to, yet. The mind naturally attempts to eliminate the alien memory entirely afterwards or at least overwrite it until it no longer resembles the original stressful event. Sometimes though it still leaks out.

These alien beings are most likely many thousands of years more advanced than the human race. We have developed most of our technical expertise in the last few hundred years and our advanced technology in the last few decades. They may even be hundreds of thousands of years ahead of us in their physics, biology, psychology and just about everything else. Nothing is potentially out of reach for them. We just have to keep an open mind, look at the evidence, juggle concepts and ideas and attempt to figure out what is going on here.

Every human being sleeps for at least one third of their lives. How many dreams do you remember? What do we dream? Could it be that we all dream some of the same things sometimes? We all then seem to forget our dreams and move into our normal waking reality. If advanced aliens have been monitoring the human race for thousands of years, they would know how to easily take control of our minds in the dream state, injecting information and direction and also know how to cover up any memory of all this afterwords. Even if we did break free of the amnesia and remember part of the interaction, we would think it was just a strange dream or collage of broken images and dream fantasies. But relax, everyone may not be subject to this interaction. Those who are may be unaware and those who are not, just dream.

I believe that we are totally conscious when we sleep but instead of being outwardly conscious we are inwardly conscious. Sleep, it is commonly agreed upon consists of five different states or stages. Stage one is light sleep where we could be stirred easily to consciousness. In stage two, three and four, we fall deeper into sleep and are much less easily aroused. Then we spend around 30 percent of our sleep time in REM sleep where we breathe more rapidly, our eyes jerk back and forth, our heart rate increases, our blood pressure rises and males develop penis erections. But the most exciting thing is that persons awakened during REM sleep report what seems like bizarre and illogical dream events! And males report bizarre and illogical dream events with erections. Why is this? Could this be a way to deal with things we subconsciously experience but can not grasp onto in our three dimensional linear version of reality? I believe this to be true only because of my own personal experiences and experiments. I believe in sleep we essentially meditate and give our conscious thoughts a rest. And then "it's play time" when we can explore. Some who practice and learn to control the "dream state" or even the meditative state claim that they can travel to whatever place they desire in their minds or work on problems they could not easily solve during the waking state where they would be challenged with so much more data input and issues as they live their daily lives. It takes a lot of practice but many people have learned that they can control the dream state and use it to their advantage. Those of us aroused after sleeping more than a few minutes exhibit amnesia. We don't remember anything about falling asleep or anything afterward. Is it that we enter a world which is not governed by our four dimensions of space-time, and it is difficult for our minds to function in such a world still tied to our normal reality, and so we also forget about it upon returning? If there really are eleven dimensions as top physicists now believe there are, could this be the place where we experience some of them? Practice time maybe? Preparation perhaps? The bigger question, the one which no one asks is, "Why do we accept this dream place at all?"

The dream world is so alien to our usual state of being, our ordinary reality, that it is completely opposed to everything we hold to be true and natural in our world. The time does not flow correctly! The physical reality changes upon a whim! Solid matter becomes putty and plastic to our perceptions! Somehow we have learned, have been told or have experienced that the way things work in the dream state is acceptable. It's the dream status quo. Multiple changing dimensions of reality, space, time and thought, for instance, are OK. And furthermore, we have learned that we are safe there for the most part. If anything happens which would ultimately harm us, we can return instantly to our physical four dimensional reality. Bang! We have all had bad dreams. We have all escaped unscathed. We have all exited at about the correct time. How would we know to do this? How? Is something or someone guiding us? Why do we all block it out upon returning? Why is it all mostly invisible to us upon awakening? The amnesia again! A fugue! Why? And again if it's a natural state for human beings it would not be difficult for an advanced extraterrestrial race after studying us for some time to determine that this effect exists and how to inject themselves or their wishes into it. It would also not be difficult for them to use the time to insert knowledge or inspiration and then erase all memory of the alien prompted events afterwards and just let the information melt into our subconscious. But what if the "dream state" *is* actually a camouflaged reality! Real in another sense which right now escapes our physical vision of our reality. An invisible reality. We have all been there! We have all seen some of the same things! We have all seen the good and bad things! But we all forget or if not we all somehow accept that our remembrances are nothing more than just a dream fantasy because upon awakening we return to our normal reality or dimensional set. The other side *must* be an illusion. Could it be that alien places or ideas are somehow mixed with dream fantasy? How long has this been going on? A day or year? A decade? Thousands of years?

All this dark speculation aside, however it evolved, I believe that dream time is some kind of positive experience for all of us. If aliens did this to us, it wasn't to destroy us or overrun our planet. If so they would have used other means thousands of years ago. Most of us enjoy our dreams and even though we rarely remember them, wake from the experience with no negative lasting effects, none the less.

OK. So you're absolutely right! I am mixing the definitions of psychological camouflage, dreams, amnesia and invisibility and injecting other concepts and you are all confused and mixed up, and that's exactly where you need to be to understand this thing! I am purposefully skirting the edges of direct involvement. I am getting things all tangled up together so that when we start discussing the photo you will be as open and prepared as possible. Seeing this being and discussing my interactions is going to take a huge leap of open mindedness for many people. Ask yourself right now. "Do I feel OK?" If you can not even seem to grasp what I am saying or if your brain hurts when you try, you may not be. If you already feel panicky or if you can not accept the thought that you are moving towards a new reality, stop. I have talked to people, and I have shown the images to people who after freezing momentarily, immediately respond that they can not see a thing. Not a damn thing! Nothing! They become angry. They completely blank out! No highlights, no outlines or anything else helps! Really! And you may be one of them. If you only want to see light and grass and get angry and expect to reject the whole idea, you are probably already psychologically tuned to *stay clear* of these beings for very real programmed psychological reasons! And so be it you will probably never see them. Go away... forget about it and be at peace. Thank you and please stop reading now! *Goodbye!* I do not want anyone shocked, panicked or psychologically damaged in any way.

Leave now or continue.   Your choice

You've been warned.

Chapter 8

Grounded Here On Planet Earth

I take it you are still here. The group is getting smaller. The photograph is (still) real. Hasn't changed.

I am only an amateur physicist. Most of my schooling and training was in advanced classes in school, articles I have read and books on everything from Einstein's relativity to Stephan Hawking's wormholes. I've read through Brian Green's "The Elegant Universe" which includes explanations of M theory and Michio Kaku's books" Parallel Worlds" and "Physics of the Impossible", amongst others. OK. So I'm a just an amateur science nerd and a physics geek! I have never even attempted to do the higher math behind the science and enormously respect those who can. But I do at least try to understand and am able to discuss some of the theories abounding in this age of constant change and discovery. I feel that if a respected prominent scientist has taken the time to write a book explaining his work in areas of advanced science, I should at least take the time to try to understand it. I also realize that beyond what the top physicists make available to the general public there must be a vast, strange, unknown world which they only discuss amongst themselves. Beyond that I fact check *everything*. I try to stay up on physics, ancient alien astronauts and UFO news and try to get as much as I can, of what I write correct. I cross reference everything. I feel as though I'm in a whirlwind. My mind has been through many wormholes so to speak, and I'm not certain it's ever coming back.

Aside from that, I believe I have a great deal of common sense. I am naturally skeptical and am not too easily fooled. I ask a lot of questions. A lot of questions. I may not have all the answers, but when confronted with an unusual event and given a logical sounding explanation, I rarely take much stock in it unless after several days of deep thought, research and experiments into the physical aspects and ramifications of such, I may tentatively admit I might possibly believe in the theory purported to be correct. I am also quick to admit when I am wrong or need to change my theories when they are proven incorrect. Having learned of many advanced theories of the nature of the universe I believe that many of the most bizarre and wonderful imaginings of the human race may actually prove possible. Now I do propose a couple of strange theories of my own and sometimes open up areas rarely seen in the light of day, but I hope that you will not take my thoughts any less seriously than more established experts because some of theirs are much weirder and stranger than mine will ever be.

Chapter 9

Into Uncharted Territory

A re there more dimensions than what we see in our everyday lives? Are length, width, height and possibly time, all that there is for us to grasp onto here in this universe of ours or are there more places to explore? Are there dimensions where our limited reality ceases to function correctly, or possibly, where our minds run wild? Could there be places where many more dimensions all interact with our senses at once and where we in turn, can interact with them? I believe that there may be, and some physicists now claim that there are. I also believe that most people are trained from birth to disregard these extra dimensions because they are inconsequential to, or do not gently mix with our everyday lives. Hell, I'll bet you right now that you don't even notice most of what is actually going on in our own standard set of dimensions let alone in additional or skewed ones. We humans block out anything that we don't need to focus on at the moment. Imagine walking down the street, through a crowd at rush hour and not noticing an ant crawling across the sidewalk. The ant is very much there and very aware of the people around it, but unless you stop and peer straight down at the sidewalk, focus your eyes on that tiny black dot, clear your senses of everything else pressing in around you, you won't even know it's there. OK. So it's a very small insect, and people are comparatively large, but people tend to not see what is not important to them or seems trivial at the moment. Do you stop to notice each blade of grass on the lawn in the park where you just sat down for lunch, and do you notice each tree in the field and what shade of green the leaves are as you sit and eat? Maybe you do, but most people are more concerned with watching for their friend who was invited to share the meal but has not yet arrived, or that moist piece of chocolate cake with the smooth rich frosting sitting waiting for them once they finish their turkey sandwich. If a stray dog runs toward you, you immediately notice it because it might be aggressive and bite. Suddenly your friend is no longer as important. The grass is still there, but the color is no longer of consequence as the dog running across it might leap onto your blanket and eat your turkey sandwich. It might even be so bold as to steal your chocolate cake! That's the limited way we see our reality. Levels of priorities, needs, interests, noise. We continually shuffle them placing the most pressing important ones on top of the pile for our brains to monitor. The lower an item falls on the scale the more likely we are to never consciously notice it.

So what I'm trying to get at is that unless you are looking for something like I have found, you will never see it. It will remain lost in the background noise forever.

Background noise. This is the stuff which things hide in! Your keys hide in the background noise of your house, in amongst the clutter on your couch. Your cat hides there too when he does not want to be seen, in the background noise created by the plants on your windowsill. Your child hides his voice there when he does not want you to hear that he is going to the playground with his friend that you don't like as he calls out just loudly enough to be heard, but competing with the television station you have on as the TV talk show host calls out her latest guest and the audience screams in delight. What did your son say? Sure it's alright. It's the new idea or inspiration hidden in the background of the dream of your first love, hidden, waiting, invisible. In the background noise. Camouflaged. As you see, by background noise, I am not referring literally to a sound obscuring something you are attempting to hear clearly. I am using the term in multiple ways or dimensions within our language. Maybe I didn't need to tell you that and maybe you've already figured that out. Background noise. Multiple dimensions. Why?

Aliens think in multiple dimensions.

The background noise can be as simple as grass, sticks, twigs, leaves and shadows. As simple as truck sounds, paving equipment, an ambulance siren, a child crying, a car screeching to a stop or the ice cream truck jingle, jingle, jingle. The whirring noise of the refrigerator slowly cycling through it's cooling period. The light glinting off the rear view mirror of a jeep, the splash of sunlight shining through morning fog or warm water running rapidly over rocks and small river stones or over your fingers as you sleep. Within the strange landscape of a snowy dreamland. Easy, yes, easy enough for any four legged creature to hide amongst or just about any other suitably adapted highly intelligent and highly evolved bipedal alien being to stand in, in plain sight.

Aliens think in multiple dimensions.

Possibly as you grew up you played with a ball when you were a child.

Later you grew to understand that shooting a marble across the floor you could hit another.

Later yet your uncle brought you a game called checkers.

You were given a monopoly game and found that it was not hard to excel and win.

One day a friend at school taught you to play your first game of chess.

Juggling multiple classes at school and excelling at exams became second nature.

Three dimensional chess boggled your mind for a few days and then became second nature.

Lectures on string theory.

Give it a few hundred thousand years.

Aliens think in multiple dimensions. Because that's what they like to do I suppose.

Chapter 10

Stealth and Magic

And so this brings us to the point where you can now clearly see that an intelligent being utilizing both physical and psychological camouflage could easily be disguised as background noise or objects which do not draw our attention so that they could operate in complete invisibility mode. They could also operate in a way that we would dismiss or forget later on in order to cover their interactions with us. They may also be using additional dimensions in a way we have yet to comprehend. This would take either a great deal of technical expertise or a very advanced evolutionary skill based upon manipulation of dimensions within their reality. Possibly our dimensional set is so flexible to them that they only need to use one or two at a time. They may be a million or more years more advanced than we are. It could easily be that they really are from another slightly different set of dimensions and/or frequencies of energy/matter, and so they can move very easily among us and hop in and out of our reality naturally. We as physical human beings are bound to this earth and unless we use our technology we are unable to take to the air and fly. But a bird naturally flaps its wings and soars into the air with no knowledge of planes or helicopters whatsoever and so the aliens may move effortlessly around us at possibly blinding speeds and levels of invisibility or just hide in the background noise with no knowledge of advanced technology whatsoever. They may, if they have traveled across the vast distances between stars, have developed star drives or methods of inter-dimensional travel before we were able as a species to walk upright or use tools. We do not yet know. Keep an open mind.

Now just imagine that possibly they have both natural ability plus millions of years of advanced technology. The result would be something not readily understandable to us at all. As Arthur C. Clarke has said, "Any sufficiently advanced technology is indistinguishable from magic." It might take centuries of study and experimentation to even grasp some of the basic information observed in their actions or behavior. It may ultimately be impossible for us as a species to grasp their abilities due to the fact that they are of another form of life/matter or energy and just using their natural abilities so alien to us.

It may all just appear to us as magic. Magic in the background noise. But if photographed it might be a revelation. An ah-ha moment, an epiphany............Here we go.............. Deeper.

Chapter 11

Arrival of The Gooey Light

Greggor and I were bored that day. I had cut the grass in the front yard and trimmed the hedges in the morning. We left for lunch in my pick-up truck and headed for the local fast food burger place. With the window open on the passenger side, Greggor would hang his head out of the truck as so many dogs love to do and feel the wind in his face and the fine smells of the city in his nostrils. He looked more like a dust mop sticking out of the passenger side window than the utterly happy dog that he was.

Oh the pleasure of a ride through the countryside for a fast food lunch. It's a dog's life. As I slowed for a signal light a neighbor waved and smiled, but a glint of sharp sunlight bounced off the mirror or maybe the trim of another car, and as I looked keenly to see who exactly was waving, it hit me square in the eyes and temporarily blinded me. I had a very strange feeling. Just for a second mind you, but it took me some time afterwards to work out just who had waved as I was negotiating the curve of the road past the signal. The light hitting my eyes also gave me a shot of anxiety, but I passed it off quickly as nothing but the fear of running into something around the corner or possibly an impending migraine which always makes my eyes light sensitive and makes me feel a bit anxious and irritable. Why is it that, on some summer days, the light is just right so that it reflects off every shiny surface right into your face. I don't know, and I suspect it's just one of those angle of the sun things, but it doesn't seem to happen to me every day at the same time throughout the year. And sometimes it's really strong.

We pulled into the parking lot of the fast food restaurant, and I decided after a second or two that the drive-through would do. Greggor wasn't dressed to go inside anyway and seconded my decision with a smart woof. I peered at the plastic menu board for a moment and ordered the classic big burger. And Greggor, well he likes breakfast food best, so I also ordered an egg and cheese sandwich for him. After paying for our meal and collecting it in its paper bag, we drove to the far corner of the parking lot and turned off the engine to eat. Greggor was especially hungry, and I wound up walking back inside to get him another breakfast sandwich almost immediately. He ate his second sandwich just as quickly, and after a short doggie walk around the parking lot we headed back home.

Another short drive and we were back at the house. Greggor jumped to the pavement as the door opened and ran towards the backyard fence where he knew his favorite tennis ball always lay waiting in the grass. It was our routine to play a few minutes of fetch after eating lunch and so I walked to my spot and he to his where he picked up the ball and stood waiting for me to signal him over to give it up into my hand. I gave the call and to my surprise, nothing. He did not move an inch. Now I knew this dog desperately wanted to play ball. He always wanted to play ball. It was his most favorite thing to do in the entire world. I call again, and he drops the ball. Dropping the ball is his way of telling me that the game is over, and we're not playing anymore. He then lets out a woof, turns in a circle and sits down. OK. I wonder, "What have I done wrong this time?" But Greggor, look at Greggor! He looks at me and then stands up and picks up the ball! I call again and he then responds bringing the ball to my side. Tease! Or so I thought. A few throws, chases, catches and returns and he seemed to be himself again. Everything was fine, though as I was playing I became aware of the changing light around me. I thought that I was not focusing well on the game or something as one object seemed to momentarily blur into another. I stood still again, and yes, all was well. It was a beautiful sunny summer day, and everything was lit with a sparkly bright amber yellow light. Hold on. There was that feeling again. It was even stronger now. Something wasn't quite right. The light. The shine. Greggor. Everything seemed too big! Too bright! Just a little too shiny!

Have you ever looked up at the moon on a summer night. The weather is warm and inviting. Your love is by your side and glass of red wine is in your hand. You gaze up at the moon, and as it rises above the horizon, four times as large as you've ever seen it before, you turn your head and whisper, but hey! Hold on, wait! Hey! What's up with the moon? It's four times as large as it should be! And further more it's redder and stranger looking than you've ever seen it. It rises, and as it does it moves slowly across the sky and through the course of the night returns to it's original size if not original color. What!!?

"Why?" You ask. Well it's a mystery. They used to say that the length of the earth's atmosphere acts as a lens to make the moon look larger than it really is but that was proven to be false. The current thinking is that it's purely a psychological effect. Yep. You heard me right. It's all in your mind! The experts insist that they know why it looks this way. But now you've looked at the moon. You've seen the size compared to so many things around it and can decide for yourself. But the experts have spoken. So repeat after me. It's all in your mind. Our best scientists just don't get it and seem to have fallen back on this one explanation. It's all in your mind. Or is it? Some people think that even though we do not know exactly what is at play here there is something beyond, well beyond a simple psychological effect. It's obvious to anyone who looks up at the moon on such a night. Now, everyone agrees that the moon is not moving that much closer and farther away from the earth. And I'm not saying that the moon physically changes it's size or it's mass. But if our own moon can play tricks on us we can not explain or understand, what could a million year old race do with all their knowledge of life, gravity, electromagnetism, magic, psychology, visual trickery, technology and camouflage do? I don't even want to think about that anymore, right now. At this point, I really just hope they're friendly. Anyway, if you hadn't ever seen it before, seeing the moon four times larger than usual, has to be a very strange experience. Bewildering.

So there I am. I'm standing in the yard, and I'm thinking to myself that things do not look quite right. At times, the grass, the trees, and even Greggor don't look like they should. Am I having heat stroke or a seizure? Nope, too young, at least according to my doctor. Have I ingested some sort of poison or drug? Ergot in my Rye? I suppose the fast food place or the CIA could have slipped some LSD into my lunch, but it didn't feel like I was tripping. It seemed as though the light, air, scenery, whatever was somehow wrong. I was certain it wasn't just me, so I went into the house and grabbed a camera which I have always kept near the door for spontaneous snapshots though I am not a great photographer and have no special skills in that area whatsoever. I am lucky to be able to point and shoot. The only camera that I had happened to be a simple disposable model. The ones you used to buy at the local grocery store that came sealed in plastic, fully loaded with film and batteries and all you need to do to use it is to unwrap, wind, point and shoot. Hello? A drunken monkey with one eye and a hurt paw could do it and not screw it up. This is not the latest digital version if you haven't guessed by the way, it's the old school film loaded mechanical version. And this should actually put some of you just a little bit more at ease. Think of it! It means that there are actually negatives. Yes, negatives! Those old, odd, dark, backwards film things that show exactly what was in a photograph before the film was even printed. And yes, I do have the negatives. And no, there have been no alterations or anything else possible (is there anything else possible?) from the processing of the negatives to the prints or anywhere else in between since the shutter was snapped.

I took aim at Greggor and the yard and prepared to shoot. I did not see anything unusual, and I was beginning to think that I really was crazy and in serious need of a happy shot, but I kept looking as something else seemed to be drawing my attention. Greggor was usually the subject of most of my photos. He was my best friend and seemed to understand people and situations in a way I would only later catch onto. He was also gregarious and outgoing, so if he was with me, which he usually was, he was always the star, and he knew it too. But Greggor was not my focus right now for some reason. I took a deep breath and looked around again. Greggor let out another short uncharacteristic woof, but I ignored him as I was then looking in the opposite direction.

When I turned back and looked over the yard again I still found no obvious weird things going on and then for some reason I just decided to let go and relax. I let my mind drop down into a much more calm mode and almost unconsciously scanned the yard sensing the brightly lit and dark areas. I realized I was looking for a point to focus on. I remembered this feeling as I had done something similar to this since childhood but had neglected the talent for many years. When I was about eight or nine, I remember letting my mind go blank and concentrating my eyes on one single point on the wall. One single point in space and time. One single point of light. That space would then expand, and I imagined that I could crawl into it. I could not then comprehend the place it created for me in my mind, but I played in that space on days when events caused me to be afraid of the outside world or when people were, in my mind cruel to me. I also went there many times because the light was so friendly and comforting. It was warm there, and I believe there were things to do and experience though most I can not remember. But I always came back with the same thought, that I was welcome there, and it was a place I wished to go back to time and time again. And now it was back as if I were there once more. There was that light and the light seemed to take on a gooey unfocused quality, nebulous and hard to define, though I tried to convince myself that it was probably all in my imagination. Not a single thing seemed like it was out of place, but there was that feeling, and Greggor sensed it as well, I'm sure of it. The woof. The spin. The strange happy actions. I'm sure that there was something going on, hold on, the gooey light! The light was just strange and different. Yes, different is the exact word. It was not threatening at all, but it was different and a bit odd. I was feeling a bit confused now. Something was there. Nebulous, mysterious, there, with Greggor. He immediately rose and sprang to life moving towards me. Surprised, I backed up a bit and to the right and froze. I suddenly needed to bolt. The urge was so strong that I felt I would vomit, but I managed to relax again. I let him move towards me and noticed to my surprise that he carried his ball. This is important and may not sound like much to you, but this dog loved his tennis ball. He would not carry that ball unless he was confident and in control, and he was carrying that ball. I believe he was saying to me, "Stay where you are. It's OK. Be at ease, relax, all is well" He approached me and moved within four or five feet. He sat. He sat and dropped the ball. It freaked me out because this dog did not normally do this until our game was over and he was no longer interested in me throwing the ball. I was very confused. We had not played long at all. Very strange. OK. I did not know what was going on. Thinking back, were the aliens communicating with him? Were they controlling his mind? Or mine? Giving him some kind of knowledge that he needed to manipulate me into position? How did he know that I needed to be where I was and calm now? Was there telepathic knowledge flowing from his mind to mine? I only know that when I needed to take the photo..... I felt like I had to take the photo..... And I did.

Snap!

The photos captured are unlike anything I have ever seen in my entire life.

Chapter 12

Return Mail From Another World

The level of camouflage which this being utilizes is nothing short of astounding. It's talents range from simulating physical aspects of the area around it, to the physiological. If there hadn't been a series of odd events leading up to the day the picture was taken, I wouldn't have ever known that the being had posed for my photograph. I probably wouldn't have known that he was there in the picture at all. I'm also not certain I would have ever noticed it. Afterwords I almost forgot about it more than once telling myself that it just couldn't be real. But upon finally accepting that it was real I began to discover more and more detail of his camouflage simulating the physical look and characteristics of everything in sight from grass, twigs, foggy light effects and leaves to psychological intimidation with weird objects and various scary subterfuge all combined with transparency and speed of movement. Aside from that, I wouldn't even have had a chance of seeing him unless my orientation to his was exactly perfect. But it was. He obviously wanted me to photograph him. And in detail, detail that has never been seen on this planet before, because here is an intelligent entity displaying some of his technical expertise, his culture, his dress, his physical form in entirety and possibly his method of gathering food or doing experiments. Just enough to give me a clear view without appearing too openly. Smart alien.

Do you recall the Pioneer and Voyager space probes? Starting in 1972 with Pioneer 10 and then again in 1973 with Pioneer 11 humans sent out a message that we were ready to communicate with and meet other beings from outside our world. The main purpose of the Pioneer mission was to explore the solar system and most importantly the planet Jupiter and its moons, Callisto, Ganymede, Io and Europa. Saturn was also explored in some detail by Pioneer 11. Photographs, as well as scientific readings, were made of Jupiter and Saturn and also of scientifically interesting places and other planets along the way. In addition, the probes carried a very special plaque designed by Carl Sagan of Cornell University bolted to the exterior of the spacecraft which not only depicts the spacecraft itself but also depicts a naked man and woman and a star map showing directions to their home, in other words the whereabouts of planet earth. It was specifically designed to be viewed by extraterrestrials.

Voyager 1 and Voyager 2, launched in 1977, contained a more detailed message which was also carried out into space. It's the story of our world on a record, a metal plated disc, 12 inches in diameter, containing images and sounds of all the diversity of life and culture on Earth. The content was again selected by Carl Sagan and his associates who chose some 120 images and a variety of natural earthly sounds. The sounds included the vocalizations of whales, birds, and other mammals as well as spoken greetings in fifty-five different languages including Sumerian, Aramaic, Hebrew, Spanish and Urdu. It's an earthly chorus of people speaking the most widely used or most ancient languages on earth relating their greetings to aliens and with the visual information on the disk, telling them of our world, how to get here and what we are all about on our wonderful little blue planet.

Now here's the interesting part. The plaques and discs on both missions were made of gold. Gold anodized aluminum and copper to be exact. The plaques, once again have engraved upon them a sketch of a man and a woman showing our specie's bodily design and external reproductive organs as well as a star map to earth. The discs have a photographic and audio record of life on our planet as well as our cultural and vocal diversity. These items were designed to teach any other race that encounters it a very general overview of us. Not much more than a snapshot but enough to let them know that we are here, intelligent and technologically sophisticated. We are sending a message, and we are friendly. Surprisingly, the probes had no landing gear! They were never designed to and were not meant to land on any planetary or solid surface. Why would you place a plaque on a spacecraft which can not land and so no one can read it? It was known at the time that the probes would never land safely, and they hoped that if aliens found them, these extraterrestrials would have a means to manipulate and view the craft anyway. The plaques exterior surface was made of shiny gold. Gold just might look very interesting to aliens and also would not deteriorate during the craft's long voyage through space so far from earth. Voyager's disks would undoubtedly be a point of interest to any alien observing them. They would be a big bright shiny sign saying, "Look Here! Read me!"All information on both these records was in visual, audio or analog form. Very little binary language was used except in attempting to relate physics and mathematical concepts.

The Pioneer 10 probe alone has now traveled over eight billion miles. It is now so far away from the earth that we have almost lost track of it and have not received a signal in over ten years. Pioneer 11 was last heard from in 1995. Are they still operational? Are they still moving? They were specifically designed to be seen out there. We know where they should be. Pioneer 10 was last seen heading out into deep space towards a star called Aldebaran some sixty eight light years away. Pioneer 11 was last seen heading out of the solar system in the opposite direction towards the center of the known galaxy. The plaque. The disc. The message, The map. Have they been seen? If the crafts were discovered by an intelligent life form from out there, they would have already been tracked and studied. Attempts would have been made to reach them, board, explore, communicate with and to understand their purpose. The plaques would have been found and examined. That is unless there were some overriding concern such as concern over contamination? If aliens touched the craft, would they be susceptible to contagions which may have hitched a ride? Keep in mind that we would be very cautious on this planet as well. A craft reaching the earth from another world would be isolated, and any human contact would be by men in bio-hazard suits until we knew that there was no potential threat to life. So these aliens probably wouldn't touch the probe or take it back to their craft. They don't yet know us and should probably not trust us. If they've been around long enough to develop technology, that would be a very safe bet. They might even wonder if it would explode. These beings would instead circle around it or send their own probe to it to take photos. So how do you find our message if you don't want to disturb the craft? Easy. The Voyager craft have the disc affixed to the exterior surface, and the Pioneer probes have their golden message in plain sight attached to an antenna support just begging to be looked at in detail.

Say a probe reaches some new race of beings who are not life as we know it but are just the same truly alive and are not incredibly technologically advanced in sound reproduction. If they are so alien that they could not read or understand the scientific message, the least they would see on the Pioneer probe is a golden man and a golden woman and a golden star map pointing towards our planet earth. Maybe they know just enough to understand that message. Maybe now they would like to visit us? It would be a difficult undertaking to immediately devise a space mission, build a new craft and travel long distances to see us here on earth, but it could be done. But what if they do not need metallic craft to travel in space? Maybe they are so far advanced that they already have the technology to get them here. Maybe they travel through and utilize a different configuration of dimensions than the ones we know of. And just maybe they are also interested and willing to communicate.

Why a golden plaque? A golden record? Why is gold so special? Gold, is one of the most stable elements in the universe that we know of, so it's no wonder that Carl Sagan and the Jet Propulsion Lab chose gold as the medium to inscribe their record upon. The best we can figure is that gold was first discovered around 3100 BC. You can not destroy gold. Every bit of gold ever mined or discovered on this planet is still intact. Every last bit. Gold is quite soft in its pure form and is also the most malleable of all metals. Malleable means that it can be beaten into another shape without cracking or breaking. It is an excellent conductor of electricity and also an excellent reflector of infrared radiation blocking out heat almost entirely. Because of it's value, creating gold through chemical means has been the pursuit of men for centuries. In ancient times, this work called alchemy, was the profession or pursuit of men who were of the same spirit as the gold prospectors of the 1800s and of today. Men who labored lifetimes to find or attempt to make the smallest piece of this most precious yellow metal. Alchemists believed that they could find the fabled philosopher's stone, which was said to be able to transform pure lead into pure gold. Even the great Leonardo Da Vinci practiced a form of alchemy. Gold was also termed a noble metal by ancient alchemists. Noble in that it is the most nonreactive metal known, and will never tarnish or rust. Scientists can now create gold inside a nuclear reactor, but unfortunately the cost of doing so is far more expensive than the value of the gold produced, and will be into the foreseeable future if not forever.

Standing on our planet, looking up into the sky, golden is the color of our sun as well as the color of light rays emanating from its surface. As they pass through the earth's atmosphere and strike our planet the sky may look blue from outer space, but the light striking out towards it from the sun looks golden. Alien beings would see this too. That's assuming that the aliens viewing it see light as we do. But what if they don't. Could it be that they think all humans are gold or amber color vision oriented? We sent out our communication to them on gold.

Kings have always treasured gold, Sumerian tombs contained gold. Egyptian Pharaohs always controlled large amounts of gold. Gold is used in religious and social ceremonies such as Hindu marriages and the Chinese New Year celebration. The Dome of the Rock in Old City, Jerusalem is covered in gold. The United States once based it currency on the price of it's reserves of gold. Helen of Troy was supposedly adorned in jewels and gold. King Solomon adorned his temple in gold. The Inca chiefs prided themselves on their jewelry of gold. The Aztecs thought gold was the excrement of the gods. Jesus was supposedly gifted by the Magi with frankincense, myrrh and gold. In all our lives gold has a very special place.

Getting back to the disks, we humans have launched into space, on an interstellar space probe, a message to aliens inscribed on gold. A very valuable, non reactive, non deteriorating metal. The message is recorded using audio and analog methods and also contains visual representations of scientific information with only brief instructions in binary language. How would aliens respond? They may have no idea what our exact level of technological sophistication is except for what they see on the plaques which is visual, audio and analog. Golden.

The Pioneer Plaque

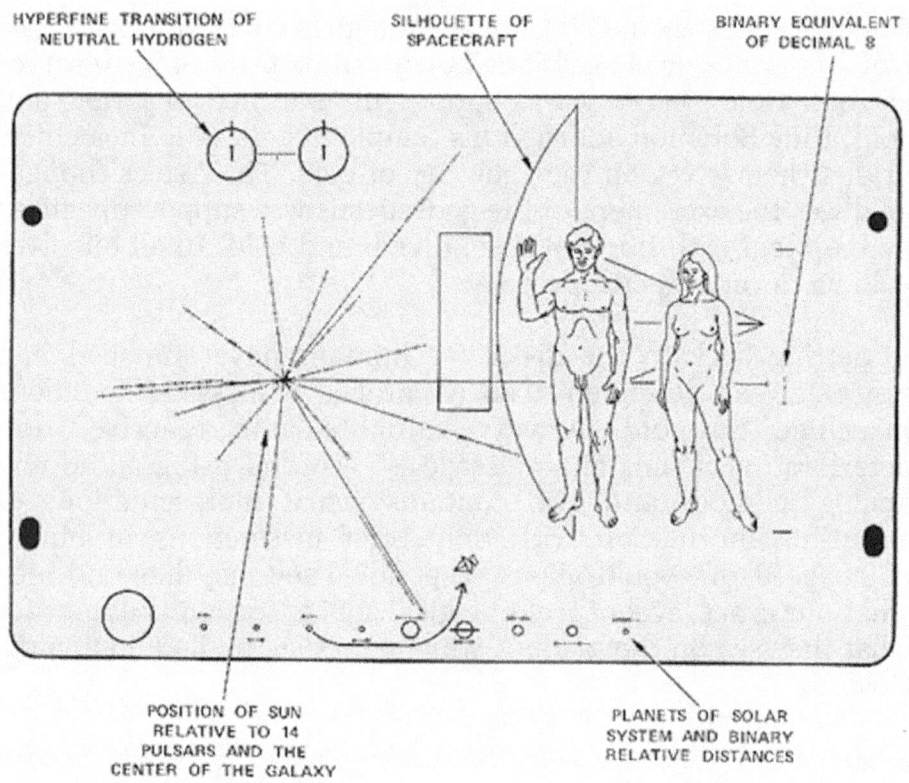

Pioneer Photos Courtesy of NASA

The Pioneer F Space Probe –Courtesy of NASA

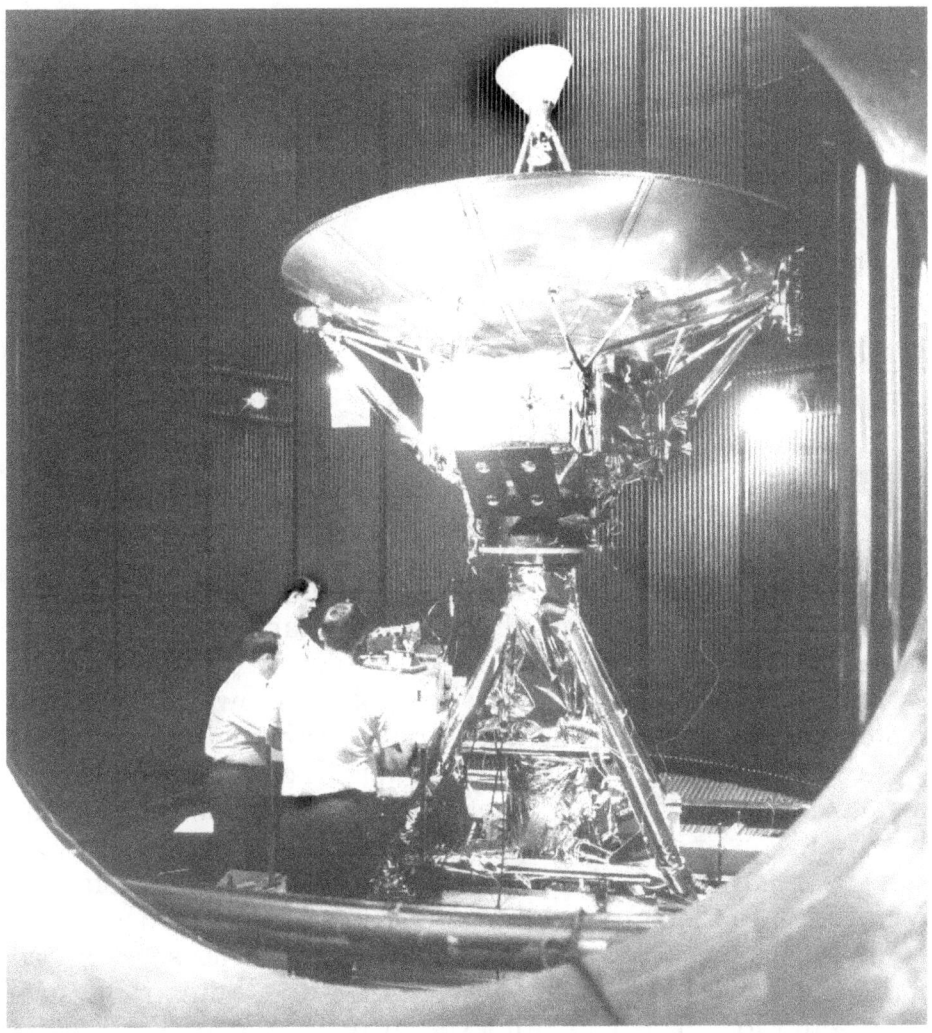

Pioneer F is the first spacecraft designed to travel into the outer solar system and operate effectively there, for possibly as long as seven years and as far from the sun as 1.5 billion miles. Its primary objective will be to take the first close-up look at the planet Jupiter, its moons and environment.

Chapter 13

The Golden Crowns of Kings

The word "crown" is derived from the Latin "corona" meaning wreath. First seen in Asia and Europe the crown has been used as a symbol of power and majesty for thousands of years. It was only to be worn by a ruler or deity. Monarchs from around the planet have worn golden crowns throughout history to designate their status and even before crowns, they wore coronas of colored leaves or flowers. A crown is a corona or an adornment of the head. Something to be wrapped around the head and yes it represents righteousness, power and glory as well as resurrection and life after death and immortality. A connection to the power of the gods. Kings wear crowns! And of course everyone pays attention to what the King says. *Every word* he says. You could say that it's the human physical representation of a halo. And yes, usually a golden halo. But why? Why would human royalty gain acceptance of their power, knowledge and authority by wearing a golden corona or crown which looks like a halo? Interestingly, corona is also the word for the outer atmospheric layer of our sun. That very active layer which can only be seen at the time of an eclipse. The part which spews large spikes of golden plasma, leaping forcefully upward and outward from the surface. By the way, the alchemist's symbol for gold was a circular ring with a point in the middle. That is also their symbol for our sun. The sun, a perfect sphere, shining with warmth, bringing daylight to the cold dark world we live in every morning. Golden. Perfect. Circular. Majestic.

In many depictions Helios, the Greek god of the sun is shown as having a halo of the sun's rays. What, you say? The Greeks had a vision of their god which showed him having the flaming rays of the sun emanating from his head? The Romans also found rays of the sun around a person's head to be important. This originates from even earlier sources, pagan sources. The Mesopotamian rulers wore a golden band around their heads. Why did the Mesopotamians feel that this was important as well? Remember in the last paragraph we discussed how crowns were always worn by two types of beings, rulers and deities. Think back to your remembrances of halos and who you remember was shown wearing one. Usually it was someone said to be a deity but often in religious art it was also someone who was under the influence of, or in league with a deity. In western history and folklore many crowns of kings were said to be offered to those on Earth by angels of the lord. The Statue of Liberty wears a crown and so do Native Indian Chiefs in the form we know as a headdress. These are all representative of a form of a halo, and a halo is equivalent to a crown.

A halo is a ring of light surrounding a person's head or representation of such. It has also been known as a nimbus or sometimes a glory. In Christian art works, it is ordinarily used to designate a holy or divine person. The same is true in Hinduism, Buddhism and ancient Greek and Roman artworks throughout the centuries. Flames, just like the ones emanating from the corona of the sun are sometimes substituted for the halo ring in Asian art works.

How many cultures have come up with this same idea for this

same thing at the same time? Why is the emanation of rays of energy from around the head so important to being a leader of the people? Whether a King or a significant religious person, one acquired a halo, corona, nimbus, glory, or crown. Why? Well you could say that they just felt that a king or someone divine looked better with his head in bright sunlight. But why wouldn't they feel just as strongly about seeing their feet pictured in water or seeing them wearing a blue coat with red dots or having long green fingernails or being naked with just a few pink flamingo feathers covering their privates? Of course all those ideas are just silly, aren't they? But why? Why are they silly but the halo is not? Think about it. Why? Because there is something compelling, subliminally acute and subconsciously powerful about the crown and halo. I suspect it has more to do with the halo, and the crown is just another physical representation of that halo / nimbus effect. Many physical objects have been used in this manner. The wreath, the band around the forehead, the ring of flowers or leaves, a golden or red disc held above or behind the head, and so there are also many ways to depict the halo in works of art. The disc or a circle floating behind or over the head, a glow shining out from the head. Rays emanating from the head, a sun colored sphere, a golden circle with spikes.

There are some definitions for halo and nimbus which I find interesting.

Two definitions of halo are:

1) A light or mist emanating from a person or object.
2) A disc or ring around the head of a divine person or monarch.

One definition for nimbus is:

--------A cloud or aura surrounding someone.

Have we all seen this thing? Have we seen this halo before and where have we seen it? Why do we immediately feel that this nimbus is correct and appropriate? Why do we all feel that this is the epitome of high consciousness as well. The essence of epiphany. The pinnacle of thought, of knowledge, the true essence of religious purity? It's the peak, the top, and if you have one, you are the man of the hour. Who or what are all these cultures imitating or alluding to? And why? Could it be that we have all seen this halo before as many of us have had contact with beings who display them? Could it be that most of us do not consciously remember this contact? But subconsciously we know. Do we all know, subconsciously that they are here? That they are powerful in ways we can only imagine? Could this information be from dreams or something like a dream in our minds? It seems we all know. You know. Can you feel it now? You know, don't you? Yes, you do.

But within about three or four minutes, you'll forget again.

That was a deep one. It takes a few to connect you with the submerged knowledge.

You have come this far, keep going.

Chapter 14

The Invisible Universe

I don't exactly remember how long it took me to take the disposable camera to the store to have the film developed. After that last time shooting pictures, I quickly forgot about the odd events and went about my days, business as usual. I was working sixty hours plus a week which left me with little time, except to do my normal household chores such as mowing the lawn and washing clothes for the next work day. After a week or two, I found the camera in my bedroom, and placing it in my pocket so that I would not forget to drop it off, I left for work. During my lunch hour, I remembered that I should take it over to the local store and hurried there. I filled out the envelope with the appropriate photo processing options and penned in my name and address, sealed it with the camera inside and dropped it into the slot as instructed. If you remember these cameras, they were totally sealed and could not be opened. Oh, I suppose if you had a hammer and a pair of pliers you might be able to open it at home, but the film would be ruined. And as I have said before, I had no expertise with photography or film developing so placing the camera in the store's processing bin was my only choice. I detached the slip with the claim number for the film on it, put it in my pocket and headed towards the door. I stopped before going out when a bottle of soda caught my eye through the glass door of the refrigerator at the check lane next to the exit. I was pretty thirsty and had not yet had time to get food or drink. I opened the door to get the soda, and as I did so I noticed there were people outside. Looking through the front door windows I could see a lady walking back and forth in front of her car. A helpful man opened the hood for her, enlightening me as to the reason for her pacing. Her car would not start. I paid for the drink and walked outside. There weren't many cars at the store and no traffic on the road. I looked across the parking lot and saw another man sitting in a car. It looked as though he also was attempting to start his car and to his dismay it wasn't cooperating either. As I was crossing the parking lot the lady's car started up, and I don't know what happened to the other man. I reached mine and opened the door. I then started the engine and quickly got back to work.

About ten days went by before I checked to see if my photos were finished. I requested only prints because I believe, at that time, there was no option to have the photos loaded onto a CD, but regardless of which, they always returned the negatives. It was Sunday, and I remembered that I needed to pick them up, and having a list to go to the store anyway I slipped the piece of paper into my pocket, grabbed my keys from the table and called to Greggor. Greggor of coarse went everywhere with me, and this ride was no exception. If I didn't take him with me when I was off work, he would likely have broken out of the house and followed me down the road! I suspect that he knew by my attire whether or not I was on my way to work. And oh yes, he was that adamant about going with me. Once I did leave him behind and was able to drive almost to the end of the block before hearing him howling like a wounded rhinoceros. Anyway, I opened the truck door, rolled down the window and Greggor jumped in. Head out the side window, he merrily road off down the street with me to the store. Arriving there I went inside to collect my photos from the bin where my name like everyone else was in alphabetical order awaiting my arrival. I found my envelope and made it through the checkout. Greggor, waiting for me in the truck sounded off as I approached. He was incredibly happy. He was usually a happy dog but barking joyfully at me on my way back to the truck was a bit weird. I opened the door and sat down in the drivers seat only to have Greggor lunge at me. As I turned sideways, all I could see was one enormous dog mouth and tongue approaching me at light speed. He licked all over my face. I tried to calm him down and after a minute or so of happy-lapping he quit and sat down on the seat and stared at me. Great! Wonderful! I reached for some napkins in the glove compartment left over from burgers I had eaten in the past. Wiping the wet slobber off my face I said something like "What the hell is up with you today?" Greggor in his usual quiet style did not answer.

We headed out onto the road and encountered some traffic on the way home. I was having a bit of trouble seeing where I was going due to the fact that the sun was once again just at that angle where it seemed to glint off of every last surface right into my eyes. I put on my sunglasses, but that really didn't help much. I was struggling to make my way safely down the road when I noticed that no one else seemed to be squinting or wearing sunglasses at all. Well for the most part. There was one old lady who looked like she had just come back from an eye doctors appointment and was wearing those wraparound dark black glasses. She probably just had her pupils dilated and was on her way home. They really shouldn't let people drive in that condition, but that's just my opinion. I managed to get to my street and pulled into the drive. We got out of the truck, and Greggor ran for the gate to the backyard. He usually wants to follow me into the house, but today, right now, he wants to play ball. And when Greggor wants to play ball, we play ball. Either that or he won't let me have a moments peace until we do. So I throw the ball bouncing it onto the fence across the yard, and he catches it in mid air, runs around a bit and returns it to me dropping it into my open hand. This is the alternate way of playing ball as his first choice is for me to throw the ball and have him chase it and catch it like a rabbit running from him and then he runs around in circles a bit and returns it to me. This goes on nonstop for about twenty five minutes or until Greggor gets tired. When he does, he starts off the same way and runs around in circles but then returns to me dropping the ball in the grass instead of in my hand.

Game over.

As I was walking to the back door of the house, I remembered the envelope with the photos and went back to pick it up off the grass, only to see Greggor sitting next to it. I patted him on the head, picked up the envelope and went into the house. Greggor followed along behind and sat down on all fours on the rug in the living room. I opened the packet carefully and looked inside. Yep. Photos. Now, I remember looking at the photos one by one. I do not recall seeing anything out of the ordinary right then. I probably looked at that photo many times without seeing, or I should say comprehending what was actually there. I'm still not certain I fully comprehend what's there now, but I've already told you that, I think. Oh well if I didn't tell you yet, I did now.

By the way. I have some strange blank spots in my memory. I thought I should say that once more now since I am sharing at the moment. I promise, I won't mention it again.

I was brought up in a nice home, the son of an advertising executive. My father, who worked at a New York advertising agency, would sometimes bring home work at night, and if I was lucky, show it to me. He was very proud of the agencies work and his own, as well. I recall that sometimes he would show me finished ad photographs. They were very interesting and I remember feeling quite proud. I recall thinking that my dad had helped to create wonderful attractive photos which helped to sell his clients products. OK. I probably didn't think it just that way, it was more likely, "Wow! Those are really cool."

One day while looking over these works he turned to me and said " There's a reason you think these photos are special." I responded that they were neat in my youthful way of expressing myself, and he listened and then thoughtfully leaned over and pointed to a place on the photo where I expected to find nothing. He said. "Just look!" I responded that I saw nothing special, with which he said, "Look again." "Here." and he took out a pencil and pointed to the spot he was referring to. He then took a few bits of newspaper and shredding them he laid them so that they covered the parts of the scene that were unimportant. I still could see nothing, and getting frustrated I stood up to leave the table, but with his large hands he motioned me to sit down again, and reassuring me that there was something to see he told me that it was a secret and only those initiated into the club could find it. Now, looking back I think he was a genius. I was suddenly transfixed. I couldn't leave now and peered into that opening in the paper laying on the photo until my eyes teared up. He then softly but in a very serious voice told me what was there. He said, "It's called a subliminal image and it's a secret way to get people's attention." "Promise never to tell anyone what I have shown you and I'll show you more!"

A mystery! A secret! I'm signed up for life! That's all my dad had to say and I'm in. All the way in and I've been in ever since. My dad was a magician! No, a wizard, like the old ancient ones in the time of King Arthur. A Merlin! An alchemist who was in amongst the chosen ones. Those who see and read between the lines. The code writers. The ad photos he showed me took some time to comprehend, and it took just as long to be able to make out the subliminal images. I strained my eyes and squinted and moved closer and farther way. I looked at them from an angle when dad said to do so, and I worked hard at it for a long time. I finally saw what he wanted me to see. It wasn't easy. It seemed like it took so long for me to see the first image. One day I looked at that photo and like a person seeing 3-d for the first time my mind went into the land of Oz. To me it was nothing short of an epiphany. There were people, objects, images and scary skulls and dead things. There were men and women together, there were gods and monsters. He said that everyone could see the people and images hidden in the photos, but almost no one could see them with their conscious mind and most would never be able to find them or remember them even if you pointed them out! That day changed my life and unknown to my father I would take that talent to a whole new level. It didn't happen all at once, but I began searching for images in every source I could find. This was my first verifiable contact with the invisible universe. It became an obsession. I spent much of my life seeing subliminal images everywhere, and I mean everywhere. After a while, I was not certain I wanted to and for a short time it troubled me so much that I shied away from public places where advertising was commonplace and prevalent. The problem was that I could see things automatically that other people could not. I began to comment on the meaning of these things and images within only to get blank stares, and many times people would move away looking confused or concerned as I attempted to relate the images, which I knew were there. Then I remembered, my father had been right. I needed to keep all this knowledge inside and never mention it to anyone again. Until now.

My photo is not a contrived or invented subliminal image developed by an expensive ad agency. But it exhibits some of the same sort of sophistication and effect as those photos. When you initially look at it, you see only a dog with his ball sitting on grass, and twigs and spots of light filtering through the trees. He is happy and looking brightly at the camera as I snap the picture.

But just as my dad said, "Look again." Here we go.

.......Deeper.

Chapter 15

The Dog With The Green Ball

A boy and his dog. A dog and his ball and a beautiful day. What more could one ask of this world? What more could this universe offer?

I hold the camera. Greggor is my subject. He sits in front of me after hurrying into position and dropping the ball on the grass. He and I both know I am not to pick it up and throw it. He has dropped it purposefully. He did not bring it back to me and place it into my hand. This is no game. No, it's not, and everything suddenly goes still. Quiet. Greggor has maneuvered me into position, and the scene is set. I recall that I am aware of the silence. The sound of which stifles every other sound in the universe so well, so distinctly that not even my own heartbeat sounded above it. Greggor smiles and sits there looking like a god damn sphinx. I know he is telling me something. I felt as though I must take this picture and I also felt an uncontrollable urge to bolt and run. I swallowed hard. I felt a bead of sweat form on my chest and in the small of my back. My forehead was already starting to drip, and I did not understand why. It wasn't that hot, but I was nervous. I raised the camera to my eye. I always center my subjects, so I wonder why I would consciously raise the center of the shot, but then as you have probably already suspected, I most likely realized what was there subconsciously. I knew I had to take the picture and the time to take it was now. The grass was mostly green in color with a few patches of tan. The weather was warm but not excessively hot, and there was no wind to speak of. There were no foreign objects on the grass, and I knew this because I would never want Greggor to have to live through having foxtails or other objects surgically removed from his paws as he needed to have done once before. I let nothing lay on the grass where we played ball. But the light looked gooey. Strange, different and somehow changing. I know this sounds weird. But the light looked as if it were changing. Foaming! Yes, foaming you might say. But then again it was just light and light is just light. Right?

Crop Circles and the forces that create them are probably important to mention at this point. Debunkers of course will tell you that all crop circles were made by Doug and Dave in England during a few short years in the 1970's. Well if those two geriatric giants of gigantic granary art forms could produce all the crop circles in all the countries, all at the same time then my dog Greggor is most likely the father of all the dust-mopey dogs in California and almost all of the cats. I say almost all the cats because he used to get tired of them after a while. He would play with mine for a few minutes letting him jump at his hind legs, chase him a few feet through his yard and sometimes wrestle with his ball. But he would eventually loose interest and so fathering hundreds of thousands of cats would be just a bit much for him. And for those of you who wish to assail me with your scientific knowledge concerning dogs fathering cats; before you do, I admit, it was a joke. Anyway, Let's get serious again. Researchers into crop circles claim that there are several different types. Two are probably the most important. The ones created by humans and the real ones.

Crop circles have been appearing as long as man has recorded strange events on this planet. The earliest crop circles were simply circles. OK. Let's take that as our starting point. What could make a circle in grain or grass several feet to several hundred feet in diameter and why a circle? Golden wheat fields. Circles appearing out of nowhere. No witnesses.

Why?

Chapter 16

Pranksters, Misguided Artists And
The Defense Of Leonardo

We all know that people imitate EVERYTHING! You can not create something unique in this world and not have someone try to imitate it, copy it or just plain steal it for their own. Everything we know of so far on our planet is made of certain simple elements. The construction of which is guided by the physical rules of our universe. It's really tough, really tough to have something happen, to record anything or see or hear something which someone could not conceivably recreate in some other way using any number of different techniques. With our simple elements and physical rules, there's only so many options. In fact, if you have enough specialists, enough money and enough time, most anything can be recreated or explained away. So given this, I have come to reject all the naysayers who claim that just because someone can recreate or imitate a unique unexplained event that there is now no other explanation than that it's hoaxed. There are groups out there who debunk supernatural claims. Now, I am a skeptic by nature and not prone to accepting bad fakes, pranks and practical jokes as reality, but these groups attempt to recreate any and all UFO or supernatural events using whatever materials and effects they have in their bag of tricks and eventually find something which will give a pretty good approximation of the actual events for their cameras. They then quite officially proclaim that this proves that the original was indeed a hoax! But what a disservice they are doing to science and everyone watching, who accept these groups as experts and respect their decisions. If you think about this critically, you quickly see that they are the hoaxers. They recreate things they encounter artificially. When they create something that approximates the original event they then claim that all similar events are hoaxed. Now perhaps only one percent of all UFO reports are real. These events need to scientifically studied with an open mind, not debunked. The debunkers are not at all certain that these events are hoaxed because it can not be proven, and they can not exactly duplicate them, but they intend to make you believe they're a hoax none the less. They create the only events which can be *proven* to be hoaxes. If they're not certain that they have made their point, they set the person who witnessed the event up for a surprise lie detector test! Now, you must know that there is no such thing as an infallible lie detector either. It has been proven unreliable time and time again. People beat them all the time. The research suggests that machines do detect lies better than chance but with significant error rates. Mostly, they will only tell you if a subject is nervous about a line of questioning. Once the test is completed, the lie detector operator must then give his subjective opinion. Digital versions are just as fallible. It's a digital throw of the dice based upon programmed parameters. You don't meet those parameters, too bad. You lose. If a person has had any type of traumatic experience, or their memory has been tainted in any way by others, or if they have been subconsciously threatened or scared, they will fall outside of the normal parameters and the operator may claim the test shows they are lying. If they have a guilt complex but are hiding nothing, and deceiving no one you may well again get a positive for lying by a lie detector operator. No matter what kind of test you take, the decision always comes down to the subjective call by the operator. Or the subjective calls pre-programmed into the software by a programmer. This being said, people do sometimes lie, and they also imitate true crop circles and thus create incredible, usually illegal and costly works of art. Which is which? The real or the imitation? Are all crop circles hoaxes? I once saw a wonderful painting called the Mona Lisa. It turned out to be a beautiful almost perfect forgery, but this by no means proves that the original painted by Leonardo Da Vinci is a fake or that Leonardo never lived.

This leaves us with the true crop circles or crop formations as they are also known.

These circles have been around since man first appeared on this planet. The most likely explanation for them is animals nesting or sleeping in grassy areas, but although this sounds logical, there is no way to prove that, in all cases, it is correct. There were no scientists around in the early days to track wildlife, take radiation readings, collect soil or DNA from the nearby brush or scat from the ground next to the circle. Swirling wind could also mat down grasses or grain to form a circle, but it would be unlikely to form a smooth regular pattern in the grain. Besides which there appears to be a strange way that the grain is compacted in some of the circles. The grain seems to have laid down on its own with no evidence of weight being placed upon it. The larger the circle or formation gets, the less likely it is that it was created by an animal or group of human hoaxers. The complexity of large circles would make hoaxing them a daunting task, to say the least, for a small group of hoaxers to make in only four hours, the accepted amount of time fields are in darkness and unattended. It is at this point where one usually begins to believe that there is something besides weather and animals involved. The grains lying smoothly, are joined by other strange effects, including some associated with extraterrestrial visits. Cameras malfunctioning. Batteries going dead when fully charged just minutes before. Cell phones loosing signal. Radiation readings falling outside typical background levels. The appearance of orbs of light flying above the fields has been reported as well as men having spontaneous erections. Deja vu!

I'm not certain just who reported that last one.

So what kind of swirling force would be required to create such a circle? Maybe something currently unknown to science on this planet. Could aliens be using their technology to make the grain swirl into complex patterns? Could creating a golden or amber circle for another alien purpose create a crop circle? Could an alien craft be involved? Could this craft rotate or cause the rotation and fall of so much grain at once? Could it be possible that the aliens piloting the craft can not directly interact with our planet without risking their very own existence? Could it be that they are much smaller than us and so are leery of direct contact? Could they naturally be difficult to see and hear, and this is the only way to get our attention? Could it be that they also have no corresponding intellectual basis to communicate with us aside from basics? Basics like a visual work of art? The basics laid out on a golden plaque. The basics sent out on a probe that was left to wander forever through interstellar space. Could it be that these basics also included the vision of a golden disc? A golden circle. A circle of swirling lines perhaps. Like the lines on a phonograph record? Maybe.

Part II

Photos and Captured Light

Chapter 17

First Impressions

L ooking at the original untouched photo, my first impression
was simply that I took a great shot of Greggor and his ball.
He sits looking at the camera, and it's easy to see that he is quite
happy to have his photo taken. As you look at him there, you can
see that his coat is easily defined against the grass, and there is
one long green blade standing up between his front legs, just
next to where his tongue is in the photograph. The direct rays
shining onto the back of his neck are clearly the light color of
summer sun and also appear as a splotch on his back just above
his tail and catching a bit of his paw. He has bright brown doggie
eyes, and his tongue is a nice light pink. His moppy coat is, well,
his moppy coat.

The grass is a mottled green of many colors, and there are
some brown acacia leaves which have fallen from the tree
nearby. Since this is a yard for animals, I have never planted a
nice sod lawn and have no intention of doing so. It's just a
playground for the dog, the cats and various other critters which
sometimes stray into the yard. The raccoons are seen most

often, every few nights or so if there's something of interest there such as cat food left out or garbage can left uncovered, but there are also rabbits, opossums, squirrels, mice and the infrequent skunk. Greggor was the only animal in the backyard that day. As far as I could tell.

You can clearly see from what direction the sun's rays are striking the grass. You can see that the streaks in the foreground of the photograph appear to travel parallel to Greggor's body. Since the light is hitting him at the base of his tail, and on his neck, this seems to be fairly obvious. Everything in the bottom of the photograph seems to be correct, except for the look and feel of the light streaks. I examined it again and noted to myself how the light looked a bit odd and suddenly remembered that feeling was the same feeling I had when taking the picture. It was a kind of a nagging difference and a bit strange. It's difficult to describe, like an unseen humming bird hovering four or five feet above your head. As I scanned over the rest of the photo, I also noticed how solid the streaks looked, but I again put this off to my imagination and reassured myself that this was just a trick of light or poor photography on my part. Deep inside though was a lingering doubt. There was almost nothing you could do incorrectly with this camera. If this disposable was not the origin of the words "point and shoot", I don't know what would have been.

Although that strange feeling was sitting there, poking at me deep inside my head, churning around, I could not find anything obviously odd with the picture, so I put it away and didn't look at it for some time. A few days passed, and I once again picked up the packet and started looking through it for different photos. The ones I wanted were photos of advertising from work. I came across Greggor's photo, and taking another look at it, I began to notice something odd in the upper left hand corner. There was what appeared to be a bright circle of light. The circle was patchy but almost perfect in form, and what's more, there was another one just above my dog's right rear leg. The lower one has a break in it so it's not quite a full circle and the grass is

clearly a different color in one spot at the very top. The more I focused on these circles the more they wrapped themselves around that odd feeling lingering in my mind. Why would the sunlight be forming circles on the ground? Optical illusion? Chance? The light in the lower section of the photo all falls in a direction consistent with the angle of the sun. I then began to see other interesting things near these points. In between the circles there is what appears to be a piece of rope, or a stick or something else, something strange. It seems to begin to curl at it's end although most of the length is fairly straight. The curled part ends in a little ring. I have said before that I don't leave things laying in the grass so Greggor, and I can play ball unhampered, and I do not recall anything being in the grass that day. I would know if a long curly rope with a ring tied to the end was floating around in the yard. Nope, not a dog toy either. Greggor's only toy was forever his ball. His little green ball.But there these things were. I felt that feeling in my mind and rejected it again. I had to be just overly concerned with the little stuff, but beyond that why was I so interested in this particular photograph? The interest slowly and silently began to transform itself into anxiety, and a feeling like you get when being on the verge of realizing that something is happening, and you know *it will happen,* but you're sure *it can't and it won't happen* even though you don't yet know what the hell it is. My head suddenly hurt. I spent most of the next few minutes preparing myself for a discovery or disappointment and desperately trying to back off to that bad imagination thing again. But I was already looking with new eyes. I couldn't stop it now, and I felt like I was about to fall. The light was just as clumpy and gooey in the top of the photo as at the bottom and more so. I could almost see it now. I was beginning to use my inner sense of skewed dimensions on this photo where there should be no reason to do so. I recalled the gooey light, the strange feeling, the urge to run and get out of there and I did not know what to do but keep looking. Then it happened. Epiphany. I blinked and looked again and here it was. Midway between the two circles of light and a bit to the right, I saw the face. I froze. I stared for minutes trying to decide if this was really what I thought it was.

No matter how many times I looked at it, it still resembled some sort of face which I had seen before. It's just below and to the right of the little ring at the end of the strange rope. It's tilted and is looking towards Greggor's tail and you can clearly see the two eye sockets, the nose and ears. There also seems to be some debris on it or possibly some sort of distortion in the air, but it's still quite clear. The chin is square and strong, and the mouth is closed and small or simply a slit. The nose is long but flat or covered by something and appears in a slightly greenish color. The right ear seems to have a large circular object hanging from it. Possibly an earring. The eyes are very deep set or at least appear that way. Just above the eyes is the straight line of what seems to the base of a headdress or possibly a crown. The whole thing hangs in mid air or mid grass as I thought at first. I suddenly realized consciously now that I was seeing the photo just like my father had taught me to see the images within an advertisement. I was looking at what I thought were subliminal images. Images placed within a photo but in a slightly skewed set of dimensions that made no sense to anyone consciously but for some reason jumps out at one's subconscious mind. And I had seen this face before, but where? I wasn't sure, but I recalled that some cultures starting in ancient times drew or carved faces of visitors from the stars. Those who they also once called their gods. How could this be connected to that thought? Something seemed familiar about it, but I couldn't place it at first. I searched my memory for answers. From which culture? From what period in time?

I quickly settled on the Mayans or Aztec's which had immediately popped into my mind. The face of the Mayan Maize God was the one that it seemed to resemble to me at first, but then I noticed that the crown is quite structured and looks as if it could be bejeweled. I convinced myself that it looked more Eastern Indian, and looked into the Hindu religion with no initial positive results so then settled again on South American civilizations for the time being. It could be a Mayan face.

Or possibly Incan?   The one they had once believed was their
creator god,   Viracocha perhaps?

Artist's impression of Incan god mask

Chapter 18

Strange Coincidences

After finding the face with its crown, my mind was reeling. I know that there's a tendency for everyone to see faces in nature. It's a well known effect, and I did not want to fall into that dark hole. Two dots for eyes and a blotch for a mouth. Maybe a rock for a nose. Everyone sees these faces, and then quickly dismisses them when they find no neck, arms, or body attached. Nothing more than a disembodied face in the bark of a tree or the coloration of a piece of toast. No details. I took another look and studied the area around the face, and as I suspected there was nothing there. No body attached to the face or anything that I could possibly imagine to be, a body. I put the photo down since I was certain that I had almost deceived myself into thinking I had found a photo of a some strange new crpto-person. That face really was weird. Something about it looked like it had been created by something besides or beyond natural coincidence. But it was all by itself. No, definitely not a person. Maybe some strange ancient artifact? No, there was nothing in the grass either.

New crpto-people! What a joke! I was certainly just seeing animals in passing clouds. There goes that mouse again, crawling across the sky. Wait a few minutes and he's gone, morphed into a huge white marshmallow. I was wasting my time and had better things to do. I put the picture away in the packet and forgot about it. I also forgot about the light, the weird events and the strange way Greggor had been behaving. For some reason, I forgot about all of them, all together, all at once and didn't bother even thinking about any of it again for months.

Work was always fast paced, to say the least, and I once again fell into that corporate zombie like mind frame, which most people do not even know they are living in until years after they are out of it. I lived, ate, and slept almost nothing but work and even on my days off, everything revolved around getting things at the house accomplished to be ready to get back to work bright and early the next morning. That summer, the hot weather, was still rolling along, on towards the fall, which most years here is just an extension of the previous season. I love these California nights, warm and dry. On the weekend, you can sit on the front lawn till 3 am and just enjoy the moon passing overhead. I had a habit of bringing out my star scope and setting it up in the yard where I would open a bottle of wine and locate stars and such in the sky. My favorite celestial body has always been the moon because of the great detail you can see. That evening I took my telescope and headed out to the backyard. I set it up and positioned it to catch the moon as it cleared the trees and the house. As I did not have to work the next day, I opened a bottle of Napa Valley Merlot and pulled up a lawn chair to wait for the magic moment when that wonderful white orb would come smiling into view. I remember that from an early age, I looked earnestly for the man in the moon, but for some reason could never find him. I kidded myself that maybe he was just tired and had gone to bed, or had moved to the dark side. It never occurred to me to ask someone where that man was or why I could not see him and it would be many years later when I finally did. Most people can perceive the face from naturally occurring light and dark areas but no, not me. I was just not put together that way. Later, in my teen years, feeling empowered by my dad's talk about images in ads I finally asked a friend who kindly drew a little circle and sketched in parts of the face. He then pointed them out to me with hand stretched up to the sky. I was amazed that anyone, not just me, but anyone at all would see this moon man who not only does not look like a whole man but also needs some work done to his face. What was everyone thinking? Craters yes, man in the moon, no. Not in my mind anyway.

So I sit enjoying the warm evening, watching the light of the moon brighten the surrounding neighborhood and slowly slink over the roof top of my house and into view. I turn my telescope and point it skyward, pull out the eyepiece from its plastic case, and slipping it gently into the socket, I prepare for an hour or two of wonder. I have looked skyward many times and enjoyed every moment, but at this point in my life I have not seen even one strange or unexplainable object, let alone a UFO. I haven't even paid much attention to the few UFO oriented television programs available on my set and didn't have much of an urge to do so. The moon seems as bright as a laser in my scope, and I am alive with my urges to explore the shapes of craters, and maybe even find new areas of interest. The evening ended when the moon had run it's course, grown shy, and had slipped into hiding behind the tree to the side of my house.

The next two weeks were a blur of activity. The car had a flat tire stranding me for an hour or so, the toilet was running, work was hectic, the house needed to be cleaned, and I wasn't getting home in time to do much of anything. I came home one evening and opened the door only to find the phone ringing in odd half beeps. Some sort of short in the line perhaps but I ran to pick it up none the less. When I did, there was no one there and the only sound I heard was a buzzing in the earpiece. I hung up, looked at the clock and noticed that it had stopped. The batteries had gone dead at 7:58 pm. Was there anything else that could go wrong? I threw my keys on the counter. It was around 8:30 pm as best I could tell, and it was Friday evening. I felt a warm fuzzy feeling coming over me as I realized I did not have to go to work again until Monday morning. I was elated. I looked in the wine rack and found a splendid Napa Valley Cabernet Sauvignon. The air in the room was quite warm that night so I opened the bottle and set it in the fridge to cool for a moment and air prior to pouring myself a glass.

I opened the back door and let Greggor into the house wagging and greeting me as he had always done. Everything was suddenly going well once again. Yes, the house was still dirty, but the psychotic phone was again on its meds and hence operating perfectly. I was lucky to have new batteries for the clock. I put them in place but had forgotten how to reset the time. Oh well. I pulled the top off the toilet and found that the stopper had swiveled itself out of place keeping it from sealing tightly. One quick twist into place, and it was good to go. No more running. I took a deep breath and began to pick up the house. I'm not a slob by any means, but when priorities are in other places, I can neglect housework with ease.

I really wanted to use my telescope that night, but since the moon was new and the sky was slightly hazy I decided that it just wasn't a go. I'd wait for another day to star gaze. Cleaning up the house was going well for an hour or so and I finished picking up clothes, throwing them into the wash basket which I left for Sunday morning. I went back to the refrigerator, and feeling the bottle of wine to see if it was cool, I realized that it was now much too cold. So I took it out and let it stand on the counter and chose a glass from the cupboard which I set next to it. The phone rings with that strange, intermittent, beeping, and I question whether I should answer it or just smash it. I look at the clock and notice that for some reason it apparently is not working again, and the time is now three hours off. I figure that it is just old batteries to blame. I hear more half beeps. I reach again for the phone. The same buzzing is all I hear, and I put it back down on the hook. I immediately pick it up again and hear nothing, no dial tone, no recording, no buzzing, nothing. I tell myself that I will call the phone company in the morning and go to the kitchen to pour my glass of wine. For some reason, the wine is once again too warm. The phone is quiet now, so I don't smash it.

I sat down on the couch and opened the book I had been reading and was attempting to concentrate on the page when I heard a strange noise coming from the back of the house. I noticed that Greggor, who was taking a nap did not stir much and so I turned my attention back to my book. Some time went by when Greggor suddenly sat up, perked his ears and twisted his head towards the backyard. I had not heard anything, but I stood up and walked into the backroom and looked out of the window. Darkness. I could not see a thing. Pitch black was more descriptive of reality that night. I went to the cupboard and grabbed a flashlight. Opening the door I noticed how warm the night was, even warmer than it was in the house. I shined the flashlight around the yard and to my surprise, nothing there, nothing out of the ordinary, but Greggor was still spooked. I came back inside, closed the door and locked it. I was not taking any chances since my dog had already set himself on alert.

It was Friday night, the world was waiting. The bars were waiting. Food was waiting, but no go. Not for me. Not tonight. I could go out, but what if someone tried to break in while I was gone. No, I'm tired and staying home. I looked back at my book sitting on the table and decided that I was also bored as hell with the novel, so I gave up and said goodnight to the main character, who did not respond. Greggor agreed that we should go to bed, and with a compliant gate headed into the bedroom and fell upon the floor. I took off my clothes, dropped them onto the chair and crawled up into bed naked. Ever since I was a kid, I have never been able to sleep well with clothes on my body. My bed, which is more of a sleeping loft with a mattress, is six feet off the floor with a short four step stairway access. I built it myself because I feel more comfortable sleeping up in the air, rather than down near the floor for some reason. Anyway, I knew instantly that I was much more than tired, I was beat.

When you go full speed all day long and then to wind down, clean the house, you're gonna crash. I knew instantly that I would be asleep in seconds after my head hit the pillow.

Sleep, but then.....

I awoke with a start, The bed was intermittently vibrating. I quickly reached over and turned on the light. Greggor was sitting, ears perked again staring towards the direction of the backyard, but my attention had been abruptly drawn to a place on the wall. There, near the ceiling was an amber spot. About the size of a golf ball, it hovered about seven inches from the ceiling and ten inches from the wall. Just as soon as I saw it and turned towards it, it started to move. At first it moved quickly to about two inches from the wall, still six or seven inches from the ceiling. I raised up on my knees on the bed and very slowly positioned my hand to try to touch it. I put my fingers within three or four inches of it. It hovered there for a couple of seconds shifting slightly side to side, then moved slowly to the wall and disappeared into it! If I had not watched this thing for some time, I would have suspected that I was having a waking dream. Now, I have never before had even the slightest, shortest waking dream but no matter, maybe that's what it was! For those of you who do not know what a waking dream is, they say it's just that. You have awakened, but you are still dreaming. Go figure. Now I have never, ever remembered having a dream of small amber orbs, but I have also never had a dream where I was attacked by swamp gas either. I know what I saw. I was wide awake after a second or two and still had time to watch this thing as it hovered nearby. It allowed me to get up and reach out towards it. It waited, probably sensing no hostile aggression on my part and then retreated to..... wherever.

I was wide awake and chock full of adrenaline now! I stood up and examined the wall where this thing had disappeared. I did not want to turn the light on, so I reached for the flashlight I keep near my bed. I felt the wall. Nothing there at all. No marks, hot spots, cold spots, nothing. Greggor was now watching my every move, and I'm not sure what he thought, but he was certainly interested. I was so perplexed that I pulled on my shorts and headed out to the living room. Greggor followed along behind. I turned on the lights. In the kitchen I found the bottle of wine, still warm, still sitting on the counter. I poured myself a glass and was about to put the rest into the refrigerator. No, I think I'll keep it out. I may need it.

I noticed that my heart was still beating pretty damn fast and hopefully the glass of wine would help. It was warm, but I finished it in one gulp. I looked at the clock only to remember that it was way off and grabbed my watch to find the correct time. My watch was working. It was 3:26 in the morning. I was still tired and wishing I was not so wide awake now, but the amber orb was turning my Friday night into a weird adventure. And my fear and amazement suddenly transformed itself into excitement. I looked around the house to see if it was still there. I tried speaking to it. OK. Yes, I attempted to speak to a little amber orb! Go ahead and laugh now. The other side of the wall was another bedroom, so I looked on the wall for marks as I had in my room. I watched Greggor to see if he could find anything which he did not. I suddenly had the thought that maybe the orb had traveled straight through all the walls, and out the other side, so moving as fast as I could, I checked the front bedroom and the outside wall as well. Nothing there. Maybe it veered off in the other direction. I ran into the living room. Nope. So I checked the window glass. Nothing.

I found myself stranded in reality, and with nothing else to do, was about to turn on the television and check out what was playing when Greggor caught my eye. He was staring towards the backdoor again. His ears perked, he was tilting his head from side to side as if listening to something which humans could not hear. Then, I felt the subtle vibration again. Greggor did not move. I listened but could hear nothing. Not a single sound emanated from the house or the yard or anywhere else for that matter. It was very, very quiet. I walked toward the back door and as I did, turned off the inside lights. I took the flashlight but intended not to use it if possible so as not to spook whatever I might find, and opened the door. Telling Greggor to stay behind because I wanted to confront this mystery by myself, I took a step onto the back porch. The night was incredibly warm and surrounded me as I left the house. I had only my shorts on, and I was still warmer than comfortable, and I would guess that the temperature was around ninety degrees, and it was very, very dark. It was a new moon and so there was no moonlight on the yard at all. The haze in the sky must have turned into high cloud cover, and there were no stars shining whatsoever. Dark is an understatement. I knew where to step since it was my backyard, but if it had not been I would have stumbled blindly into everything. And near blind I was. I walked forward. The sound of my bare footsteps on the walkway was the loudest sound in the yard. I tried to walk more quietly and slowly instead. Not an insect was stirring, there was no wind at all. Stillness gripped everything.

I cautiously walked off the brick pad and into the yard, sometimes turning to look back at the door of the house. I'm not certain why I was looking back, but I suspect I was checking to see if Greggor was alright. It didn't make sense, but not much made sense that evening. I had set the flashlight on the BBQ. I moved out further into the darkness alone.

As I walked forward I could not believe how quiet it was, and peering into the yard I wondered if I had ever seen a night like this before. Nothing was stirring. As I neared the acacia tree, I looked up and to my surprise a light was shining through the branches. Why hadn't I seen this light hitting the ground below I wondered? I looked back, and there was nothing, no trace of light on the grass. Intrigued, I moved forward. I was also attempting to sense the intermittent vibration which happened earlier, but I suspect that it only affected the house structure. I was standing underneath the tree when the night sky above and beyond the tree came into view. There in the sky was an orb. A larger orb than the one in my bedroom.

A much larger orb.

It hovered in the night sky and was around fifteen feet in diameter. Slightly longer from side to side and oval in shape, it looked a bit like a squashed egg suspended in mid air just beyond my tree. I'm not certain if it moved there as I walked underneath the tree, but I had to look out about twenty feet overhead to see it. It did not shine down any light at all. I now knew why I had not seen light on the grass. It may not have been there at that moment I walked through, and it did not shine down any light from it's surface. It just glowed. It seemed to be lighted from within. It's color was an amber white with a hint of blue, and it had a slightly darker amber area around the outside. I stood and stared at it for a few minutes. Then as I looked down again I immediately noticed that there was now a circle of light on the ground about twenty feet in front of me, and to my surprise there was a figure standing in it! I struggled to understand what was going on. It didn't make sense. Why had I not seen the light circle on the ground as I approached? Or the figure? The only thing I can think of to explain these strange incongruities in my attention and vision would be either these changes happened so fast that I was not aware of them, or I was experiencing time outside the norm. I lost track of myself. There's a blank spot in my mind and then I became aware that I was once again looking at that person there standing in the light. I strained to see who it was. It looked as if it were possibly a woman. Yes, I can suddenly see clearly, it's a woman or possibly a man with a feminine face. She did not move a muscle as if frozen in time. She was dressed all in white and did not look in my direction. She seemed to be looking at something off in the darkness to the left of me. She did not say a word. Nothing moved. I could see her whole body though no light shone on her. There was also nothing above that could shine a light down except that orb, but there seemed to be no connection between the two. The orb in the sky hung silently glowing, and the woman stood in the circle of light, and neither moved at all. Then the orb started to float towards me.

I suddenly felt really creepy. I felt as if something were in the deep darkness behind me. An imaginary creature slithered down my bare neck and back. Down around my ribs. The woman whose body seemed completely frozen then twisted just her head and looked in my direction. It was as if someone controlling a manikin had just swiveled the head around. I don't remember her actually looking right at me or seeing her eyes but suddenly I was back at the house opening the back door. I had somehow crossed the length of the yard without knowing it, and the orb was gone. I looked back at the tree, and there was no light there once again. I closed the door behind me as a shiver once again rolled across my arms and down my back. To this day I question whether it actually happened or not.

But I remember. I do remember.

Chapter 19

Strange Technology akin to Magic

The photograph. Yes, the photograph! That unearthly photo which also gave me that strange feeling! I told myself that I would come back to it, and really look it over, and see why it makes me feel so weird. And why so much time passes I do not know, but from that day I went through periods where for some reason I had no interest in it or after thinking about it I immediately lost interest in it. I just did not seem to want to deal with it at all. I didn't even think of the dam photo. I simply forgot. It's like some floating reoccurring amnesia, and I still have twitches of this phenomenon today. It just escapes my mind when I leave it for a moment. I have been through periods of months where I had only a fleeting recollection of it and periods when the thought of working on it scared me a bit. Then I forgot again. But today I sit down and recall the circles, the rope, the face and I remember everything. My head hurts. I keep going.

One day I was looking at it and noticed that the circles of light were similar to what I had seen in the back of my house. The only difference? In the back, there was a woman standing in the light, or was I dreaming. OK. I swear. I was awake! So I'm thinking, "Why are there these circles in this photograph?" If I did see the woman standing in the light which I believe I did, why would this look similar to the circles in the photograph? I started to look at the light and dark parts of this area. I then, for some strange reason decided that I must stop looking at this as a real photograph and instead envision it as a modified advertising photo. I had to stop looking for proper physical objects, and as I had learned, mentally let go and use my mind to see whatever was actually there no matter how strange or bizarre. I closed my eyes for a second, and when I opened them I immediately saw that the lower circle has a dark break in the lighter areas, right below where the Mayan face is. But Why? I recalled the woman in the backyard circle. She blocked my view of part of the circle. But the Mayan face is not attached to a body. OK, maybe that Mayan face is a diversion! So what is it that is attached to it that may be blocking the rest of the light ring? I noticed the dark patch extending down from the edge of the chin of the Mayan face. It heads down towards Greggor's leg and then splits into two pieces about half the thickness of the upper section. It hit me like a brick. Legs? Feet? But where's the head? My eyes shot up towards the upper circle. Now wait! This is a real photograph! Remember, if this is an imaginary something there shouldn't be anything really there, right? There should be no head, no shoulders, arms, or anything else. But looking at the upper circle there appeared to be another head, of sorts. I couldn't believe my own eyes! It definitely does not look like a human head with a human face, but there is a mouth and a nose of sorts. The head is lit by part of the upper circle. It seems to be peering off to one side towards the camera or to something slightly behind it so only one eye is completely visible. Seeing the head where a head should be atop what appears to be a midsection with legs and feet below in proper proportion and orientation, but in that strange advertising style subliminal camouflage world usually apparent only to our subconscious is suddenly completely mind boggling to me. But regardless of this, he or she or it is now exposed! I took a very deep breath.

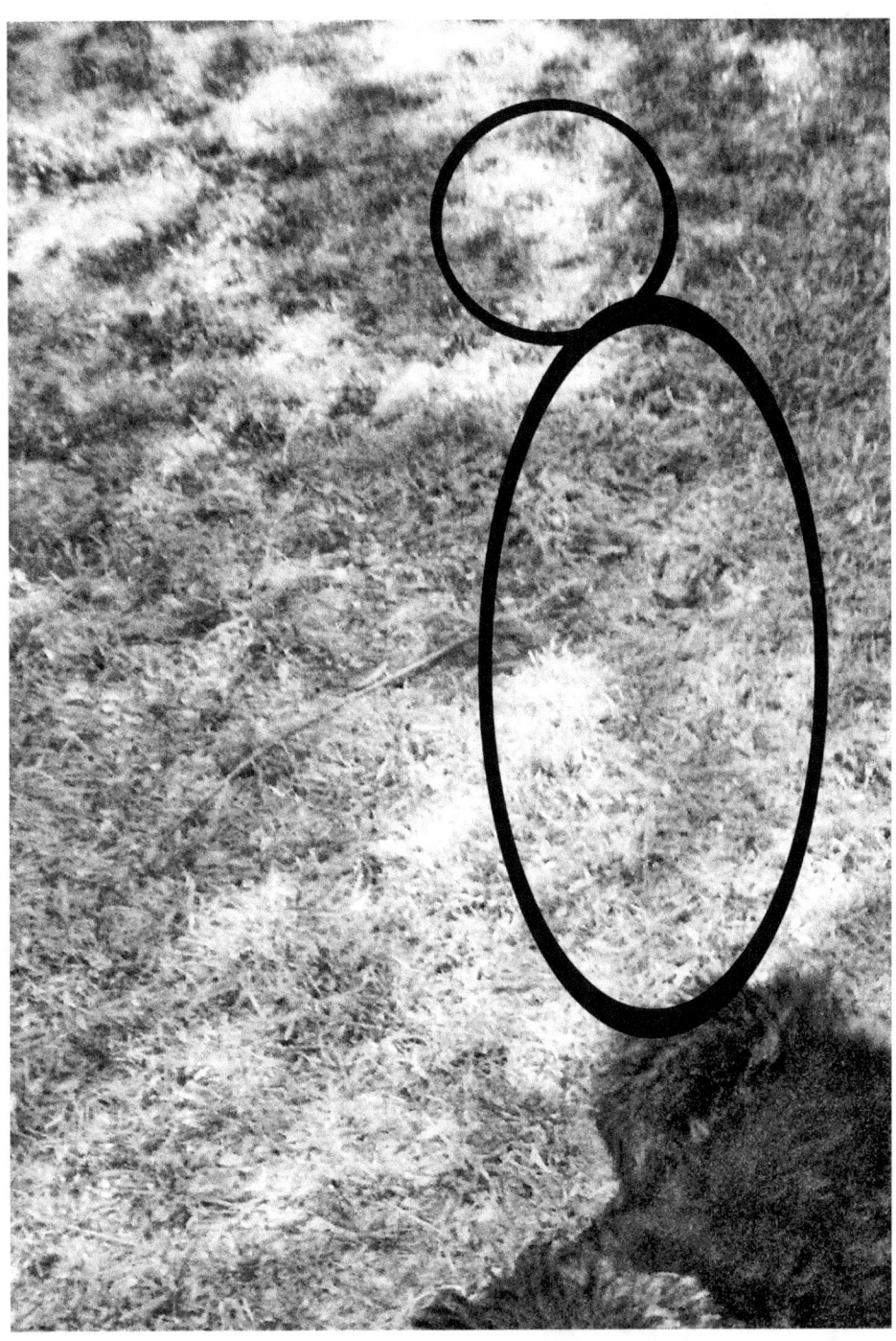

Thanks dad!

Now if you are a dyed in the wool skeptic you are probably chanting pareidolia, pareidolia, pareidolia right about now, and yelling that I am just putting bits and pieces of grass and twigs together to form something which resembles a face in my own mind. You are thinking that light and shadow could not possibly be part of the disguise of a being. Aliens could never acquire the skills to camouflage themselves to look like grass and leaves and light! Come on people! The Martians are all dead. They died from inhaling swamp gas!

I hear you loud and clear. I'll quit now with the swamp gas jokes....but the photo is not a joke or a hoax.

Pareidolia is mankind's tendency to see faces in inanimate objects. Aside from an imaginary face there may sometimes be something that seems to be a bit of detail, usually there is not. There is never any real extension of the face to neck and body parts or details of hands or fingers. Anything like that. Within a photograph (if there is one) complexity of detail is missing. You might see the general shape of a butterfly in an inkblot but there just isn't much of anything detailed whatsoever and no context in reality. Some small context might possibly be called "genius loci", which is also called the "spirit of the place" as in seeing the profile of the face of an Indian chief in a stone cliff near ancient Indian ceremonial grounds. Neither of these terms applies to my photo or experiences.

As you can see in the photo, the face is lit by bright sunlight as are the shoulders, but the body and feet are in shadow and appear as such. They are much darker (grayer) than the other body areas

Chapter 20

Epiphany and Beyond

The right eye, if indeed it is an eye, is pretty easy to see and once you connect the open slit mouth and some sort of cheek bone between the two the rest pops out at you. I'll refer to it as he for the time being even though that did not appear to be the gender of the "person" in my backyard, in human terms at least. His other eye as I have said before is mostly obscured by the bridge of the nose or angle of his head to the camera. You can see a slight indentation where the left eye socket ends and, so it's easy to conclude that there are two eyes on this being's face roughly where our own are placed. Not convinced yet? I have some additional evidence which I will tell you about a bit later which should convince you that he does have two eyes. Above the eyes is an area which is clearly his forehead. Not much else to say about it except that it separates the eyes from the top of the head. Around the head appears to be high points similar to ears on a Doberman Pincher, except, unlike the dog, there are numerous points running across the entire forehead. I count eight to ten. They look much like the glowing points of a corona and vary in size and shape. This could be the equivalent of human hair or at least something present in the same place. Where the "hair" is about the upper side of the head, the coronal points blend together into what looks like ears. Hence the Doberman Pincher reference. This could be ears with hair growing atop the head in between, or coronal structures for some purpose unknown to me. Most of the entire body is oriented in the same direction as my dog Greggor as he sits on the grass. This includes the arms, feet and torso as it would most likely be for a human being. Only the head is turned. Funny that they both have their heads turned to the left and more towards the camera than anywhere else.

Coincidence? Imagination? Come on! How much do you need to believe that this is the first true clear photo of an alien being ever taken?

My dog sits exactly the same way, facing the same direction as the alien, and his head is also turned a bit but only enough to face me. For some reason, the alien being seems to have preferred to swivel it's head a ways further. But why? Let's back up a bit. I don't want to neglect the arms, and yes there are arms protruding from the shoulders which appear to be on opposite sides of the body. Weird huh? Kind of just like a human body? Well, this gets even weirder. The arms appear to be larger at the shoulders than they are at the wrist. The hands appear to have one large thumb and a webbed finger structure. It's difficult to get an exact count, but it looks like three fingers and one thumb on each hand. I could be wrong though, it is hard to see. Pretty cool though eh? Don't ask me yet what he's holding. I won't tell you till later. Anyway, this all demonstrates bilateral symmetry. Everything about this being seems to have evolved about a central plane and has divided itself into equal mirrored parts. So people, guess what? We are most likely related in some fashion which means we are now not alone here on this planet and now have relatives coming to visit. One more time for those who didn't get the bilateral symmetry thing. This being displays identical ears, eyes, arms, hands, fingers, legs and feet on both sides of it's body. Unless nature just does this with all intelligent beings, there must be some small central imperative such as DNA or the likes to tell it to do this as it grows, which means, in one way or another..... relatives!

I can just hear it now. "Hey kids, your uncle Sparky is here from Sirius 6B to stay the weekend! He's going to need to stay in your room so fetch the tent from the garage and you can campout in the backyard for a few light years."

So once again, how much do you need to convince you that this is a truly new alien being in my photograph? We've covered the pareidolia thing. There's way too much detail in coordination with the face to be a coincidence. What about that other face, that Mayan face which I saw first? It just doesn't add up. It is in the wrong orientation to be a person's face and this being would not have another head protruding from its midsection. It just doesn't fit. Is this the fly in the ointment? I asked the same question.

Chapter 21

Red Rain

I was getting up one day and pondering this question when as I pulled on my pants and reached around to put my belt in place through the belt loops I suddenly had a bright idea. What if the "Mayan" face was not actually a face at all ? And, what if it really wasn't supposed to be attached to a body? What if it were a "Mayan" style mask? That would explain why it seems to be hanging in mid "air" or whatever with no discernible body attached. And that's what had always bothered me the most. Nothing was attached to this face. And one face with no supporting information was as good as a big red flag that all the other information was possibly coincidence. At least at this point. But what if that mask was an ornamental or decorative piece, possibly to denote some type of status? That of King or emissary! Where would one carry such an object when one's hands are full? On something tied about the waist? On a belt? So if this is true, we should be able to see this mask at precisely the point where the waist on a human would be and hanging down from there. Bingo!

At this point, my brain was leaping out of my skull. Each theorized piece was falling precisely into place. This would be why the Mayan face seems to have dark spots for eyes. Because masks have holes to see through! And the coronal hair looks more like a bejeweled crown. Because it's not really a being. It's really an ornamental mask possibly denoting high stature. Very high stature.

I suddenly had the most amazing feelings, thoughts and questions wash over me. All at once I asked myself, " Why me?" If this is a king or emissary, why had he chosen me to stand before, to appear to? The planet is filled with billions of humans, many smarter than I. Most better looking than me and some small group might even be a bit crazier. Why me? Aliens do not do things by accident. Like red rain washing over me, I knew the answer but had no proof at all. I must leave no stone unturned. No item left uncounted.

I snapped my brain back into it's sane-socket just as quickly as possible.

I took a deep breath and got back to reality checking. I did not want to hop on board this crazy train because I didn't look into all the issues. The mask is distorted on one side of the face. This could be an indication that possibly it's not really a mask, and I'm just wanting to believe so badly that I am overlooking the obvious. Of course, I'm really not wanting to believe at all at this point! If it were actually a mask, it would be nearly perfect in bilateral symmetry, just like it's owner. I was worried, but my supposition was that the mask is real, and there must be some other reason why the left edge of the face is distorted other than because it's not really a mask but a collection of weird coincidences. I needed more detail. The mask is small at best. I strained to see the edge clearly but could not see anything that would be making a negative impact on it. Options crossed my mind. Possibly an accident? A crash or fall could have dented the mask. But the distortion of the left side didn't really look like physical damage, but it's so hard to see. I needed some way to get a closer look. I headed into my room and dug into my boxes of old school supplies. I found it, a large square magnifying glass. Hurrying back to my desk I sat down and focused the lens on the mask. At first I didn't see anything but distortion, and as I sat there thinking, Greggor walked in and stuck his nose under my arm lifting my hand and changing the distance to the paper, really distorting the image. I stopped breathing. My mind went ballistic and instead of pushing Greggor away I got out of my chair, went to my knees and gave him the biggest hug he'd ever gotten. If that dog didn't know what he did, and even if he didn't care, he was completely pleased by my reaction. He jumped back and gave me a smart "woof" as if to say. " Aw shucks! It was nothing, and thanks for the hug!" I jumped up, went into the kitchen, threw him a doggie treat and dove back into work, picking up where I had left off.

The movement of the lens in my hand had distorted the image in the photograph. And what I saw when I refocused the lens was..... a lens distorting a mask. There it was! Hanging from the belt area around the beings waist was a circular lens, possibly a magnifier, and that lens just covered the left edge of the mask distorting it's image around the left eye and cheek! The lens has some sort of opaque ring around the outside and looks as if it is attached to a cylindrical handle.

Now I'm on a roll! I just had to take some time and examine the area around the mask to see if there were other objects in the same vicinity.

The long staff with ring

Looking to the left of the mask and slightly above something again caught my eye. I noticed a long thin object protruding from the belt area. It is attached to a point above the waist and warps around something there or is twisted into a coil. This is the same rope stick thing with a ring attached to it's end, which I had seen earlier. The ring turns into coils at the very end of the object. It could be a sensory device or stabilization tool such as a cane. Whatever it is, it looks like it could be coiled up to the waist on a moments notice, but when extended looks like a staff or wand. Since I know of no obvious similar human parallel, whether a device or body part, I will only say that it seems to be part of the attachments to the waist belt tool carrying area.

Do you believe yet? Are you still asking yourself if this photo is real? If it could be a hoax? What more do you need?

I've been relating the events and strange occurrences to you which I have encountered. I have also been describing the alien in detail, which you may also examine for yourself. Let's recap.

So far we have:

-- Possible mode of transportation. (the orbs)
-- An entire body representation with all parts in
        bilateral symmetrical detail. (not just a face)
-- A tool attached to the waist of the being. (the lens)
– A ceremonial, artistic, decorative object attached to the
        waist of the body. (the mask)
-- An unknown (non-human) device or body part. (the
        rope, stick, wand or staff thing)

So, more? Do you need more to make your brain click over into acceptance mode? What do you want? A sign saying, "Hello earthlings! We are here to explore your planet. Do not be afraid!" We are your friends!" Maybe that's it! Some may be rejecting the idea because this being does not seem to be alive enough. All these parts and things might just be coincidence, could still just be monkey text! They just happened to appear in what the minds eye could be appropriate places and with that, the mind then assigns proper purposes for these unrelated things which happen to look just like tools that such a being like this might possibly carry. It could all be swamp (oops! I almost said it again). It could all be a figment of my imagination. A bit of sour ham or foul pea soup consumed the night before.

Let's see. Just what might make you believe. I'm not sure these guys want to appear on late night television talk shows just yet.

So let's get back to the photo. If there were just something a little more human in that damn photograph! Something to prove that there's some sort of biological action going on. You could say that it's obviously not a real being because it has no biology, no "humanity", no life whatsoever. It could be only a ghost or something because there's only one. That's it! It's just a ghost image. It's not really alive! Ghosts just hang around, lifeless after they die and bother people. They don't really think or feel or have friends or make babies. But aliens would. Wouldn't they? Hold on, one minute. That may be it!

So, what would make you believe?

How about if he or she had..... a child? How about if this child looked almost exactly like a miniature version of the adult, just as it would with the human race? What if this child was not just hanging in mid air or off upside down in another area of the photograph somewhere? What if it is exactly where a human child might be? What if he or she or it was carrying this child? Showing it to you?

A proud parent.

And this would give us clues to their basic method of reproduction? Would that make you believe?

Though maybe they do not have genitalia that we can see on the outside, their children, could still be presented to us. We have shared images of our reproductive organs with the aliens. Remember the Pioneer and Voyager space probes? Yes! A naked man. A naked woman. The information is out there and may have been intercepted and studied and yes we may have a response! A wonderful intellectual response but not like anything we have yet to imagine or could have anticipated. So where is this child? Where is this mysterious little being which should not appear in any imaginary dreamscape? Remember, earlier in the book, I described how the alignment of the beings body is parallel to Greggor's? And that both their heads are turned slightly to the side and facing the camera. I also noted that the being appeared to be turning it's head just a bit further past perfect alignment with the camera. He seems to have something else to draw his attention to, besides me. Perhaps it's the "very human" or "very universal" tendency of a parent to turn to your child and say,

" Look at the camera sweetie and smile!"

If you haven't yet found the little rug rat, look again, for on the aliens' back, just above his shoulders you will see a head, with a childlike face compared to his, peering openly in your direction. Its arms hang onto dad's back just below the shoulders. It's legs, bent at the knees, affix themselves to dad's lower back. It's foot, the one we can see on this side of the body, hangs alongside. It's hips and butt can be clearly seen but appear to be of a different color, a whitish color. Why? Perhaps a diaper? Considering it's parent, this little guy looks exactly like what we would expect an alien child to look like. Cute too! You can almost hear him asking, "Is this one of those alien beings you said we would see? Is it friendly dad? I like it. Can I take it home with me?"

Go ahead. Take a minute here to think this one over.

The baby alien face.

The whole little alien.

In diapers..........?

Is that enough yet, to believe? You're thinking it's much easier to put it out of your mind, right? To forget?

You probably will forget in the next few minutes. I've seen it before. I'll try to bring you back again.

Oh! And, by the way! Please leave the book sitting in plain sight, right in the middle of the table.

That will help.

Chapter 22

Enigma

Once more. I have almost no artistic ability. I can not draw whatsoever. No one else has touched the photo.

The child. We were examining the small one in relationship to the adult.

Check the line of the skull on each being. The adult is looking slightly behind with head tilted back a bit, and so we can not quite see the entire face. The left eye is hidden to some extent perhaps in a blur of light, and the forehead is visible but the coronal line is not. How do we know the adult has a left eye socket or symmetrical face structure for that matter? Look again at the small one. The child is looking straight at the camera and so you can see it's forehead in entirety. It has two eyes. If you look at the adult, you will see exactly the same facial structure except that he clearly is looking up and to the left so that the top left side of the head is mostly hidden from view. The child's head is clearly in view, and so we now have, between the two beings, a complete set of features we can say are common to both.

The skull appears to be wider at the top than at the eyes or the chin, which highlights the large forehead, and yes, the eyes! It has two eyes. And just where our hairline would be is what looks like a row of ridges or spikes similar to human hair. Similar but not exactly just like human hair. The ridges or spikes also look a lot like an amber or golden colored corona. Yes! A corona, a nimbus! It reaches up with its jagged edges and circles the skull or at least as much of the skull as one can see looking from the angle presented. Is this just a hairy or bony structure or is it some form of evolutionary enhancement allowing these beings to do something which to us would only look like magic. Telepathy perhaps? There's lot's of possibilities. Let's face it, we have a photograph of a being standing in a grassy area, and it is utilizing light, energy and space-time in a way we have yet to discover. It may be even using all eleven dimensions of the universe in a way we know nothing about! Hell, it may be from another universe altogether! Let's look into some of these questions because whatever it's doing, and however it is doing it, it most likely looks more like magic to you and me than science.

A mystery. Questions and more questions. Yes, but good questions just the same. They all need to be considered. Some are pretty far out there or "crazy". Some are not. We need to understand, and we can not afford to close our minds now to the magic.

This may be the first glimpse of the man behind the curtain. The entities which may have visited here for as long as we have been here or longer and have not yet been seen or recorded. Not until now. Could they be able to subtly influence mankind? Could they be made of something as new and exotic as "dark matter"?

Now, no matter what my experiences have been, all you have to grasp onto in reality is an image on a sheet of paper. I have given you my word that this image is real and is not a hoax of any sort. What we make of the information within the image will lead humanity into a new age. Some of the speculations will prove correct and many will not. Some questions may hold truth, and many may lead to dead ends.

But move forward we must, for to dismiss and forget leads only backwards into the darkness of ignorance.

Let's explore some of the enigma presented to us. Or leap into the unknown void as the case may be.

Communication. Might the corona on the top of the head be something other than hair as on a human head? I have said that to me the spikes or appendages look more like a corona or crown. Could they have evolved as a way to communicate? A form of telepathic communication possibly? If so, they could very well be staying in contact with each other by sending messages and receiving them with their head gear arrays. Many creatures on earth use similar strategies. Some use ultra low or ultra high frequencies which are inaudible to humans as well as scents and body movements to communicate with each other. Some use chirps and clicks which blend almost perfectly into the background noise. All of which are usually invisible (inaudible) to humans. Elephants communicate over miles of terrain using very low frequency sounds below the range of human hearing. Before this was proven to be correct, they were thought to be using a sixth sense, or this communication could just as well have been called telepathic. Boney structures on the heads of aliens could be just the thing to emit, amplify or pickup low or high frequency sounds or even electromagnetic waves. It could be some sort of antennae like adaptation, to pick up and send information, in a form we can only imagine. One which our science has yet to discover. These beings that I encountered were mostly silent, and we see from the photo that their mouths may be only a thin slit. No teeth or tongue are evident, though unless I smile you wouldn't see mine either. So, this does not mean that they could not use their mouths to talk, but since we have no evidence of them speaking, or making any sounds in our earthly atmosphere or air, we might at least cease to assume that they can speak in the same way that we can. Their mouth, may be simply for ingestion of food, or vocalization of sounds or frequencies used by them for location of objects. That's assuming they have some sort of atmosphere where they come from. They might live in a hot gaseous, plasma atmosphere where we could not survive for more than a moment. Possibly if they are from a slightly skewed set of dimensions from ours, they may have no need of communication and thought just passes from one to another when needed. Hence the appendages. Spooky-com or spooky communication. So at this point,all I can tell you is that any intelligent verbal communication between them and myself has been limited to, shall we say, almost nothing. Nothing I can tell you that will make any sense. Nothing that would show that they know how to speak to us using sound waves traveling through the air as humans do.

How about travel? Could the appendages be a form of antenna that connects these beings to other dimensions and gives them access to strange and wondrous portals of travel? A path through space different and unique to their race? Perhaps it's what we call wormholes, or something so alien to us that we have yet to imagine it. The larger being is standing in a circle of light. The woman I encountered in my back yard was also standing in a circle of light. Now since there are two different looking beings but the same type of circles of light, let's assume they are doing the same thing. I further assume that they have some control over them, and they have a purpose. Otherwise, why stand is a damn light circle at all! A photo prop? Not likely. But then I'm no alien. Perhaps it's another sign of stature? Anyway, the circles which they stand in appear to be controlled by an adult in some way, but there is no control device apparent to us. Or is there? My mind keeps screaming to me that there's something hidden in plain sight. Something which we might not see as a control device but never the less most likely is. There is one other area we might consider. There's a section of the beings skin (or suit) which appears to be more active than others. I'm looking at the area on his stomach. It seems to radiate spidery colored tendrils possibly channeling some sort of force. But what kind of force? This may be a sophisticated belt which creates a field enveloping the wearer in some kind of energy, allowing them to transport from place to place. The golden rings may support this theory as they are under, and atop the being at the feet and the head while he is stationary but interestingly this is at odds with the night time encounter I had where the only ring seen was under the subject. Possibly the photo was taken just a little more quickly than he anticipated, and the ring at his head was in the process of turning sideways and eliminating itself for some reason. Stability issue in the transport system perhaps? Instability in the wormhole? The PAUSE key on his remote? Not sure but there's some great info there. The whole process could be controlled by the orb or sphere which I saw later on. Scotty controlling the transporter from the engineering bridge?

Is there something in his hands? Why yes! How observant you are. But what? You'll just have to wait a bit longer.

His hands though are quite interesting. There appears to be large, long, opposable thumbs. The thumbs are set left and right with each on the outside of the hands, with the palms facing up. From what I can tell, there are only two or three other fingers with possibly some type of webbing in between on each hand. Take a look at your own hands. Turn your palms up and touch your index and middle fingers together. Spread out the other fingers and you have something like the alien hand. This is how it looks anyway, but until you get to shake hands with one you'll just have to take my word for it. Not too unlike our own. There may also be some type of fingernails, but it's difficult to tell for sure. The hands look quite dextrous. I could see them playing a musical instrument, writing poetry or scratching their nose, all with the excellent range of movement which their arms, wrists, hands and fingers afford. Not all at the same time though. They appear to have jointed arms much as we do. They seem to use their hands much as we do. They also seem to be a fairly good model for any intelligent race of beings which have a large brain, have evolved to walk upright and have a need to manipulate a wide range of objects such as tools, with its hands.

The feet are the most difficult to describe because he is probably wearing some type of footwear. This makes perfect sense since standing and moving on unknown terrain would require some sort of protection for the feet. You never know when trekking about an unexplored planet that you might just step on a rusty nail. You don't want to have to cancel your exploration vacation and wormhole home for a tetanus shot so boots make perfect sense. You can just see a vague outline of them which could be because they are one of the few areas which does not need to be visible or be covered with anything else making them the best camouflaged area of the body. The next best place would be the legs, and you can see that they are also almost hidden in a foggy shadow and camouflage just like the "boots". Now if you look at your own feet, what do you see? They sort of look similar to your hands, right? It's that strange biological logic of bilateral semi-Fibonacci stuff we started to look into earlier. Just get a biology book and read it. I'm too busy with the aliens to explain that right now. Anyway, your feet resemble your hands. Same number of digits. Your big toe is kind of like your thumb and so on. We could, therefore, predict that these cool little guys have feet which resemble their hands in many ways. Probably not quite as dexterous, because the more you need to use your hands, the more you walk upon your feet, which tends to specialize them a bit towards balance and mobility, and so they probably work pretty much like ours. Good feet to stand on, walk on or dance upon. Not really too good to write with. Just take a look at the little one's feet. Get the idea?

Interestingly enough, the head, shoulders, arms and hands are illuminated by the upper light ring, but the legs and feet block the light of the lower ring. Why would this be? It could be that whatever device senses the background image to be displayed on the front of the suit may be aimed downward at the grass directly behind and to the right of the being! The light ring, therefore, may be much lower, and so is not being picked up by its systems. The arms, shoulders and head are at the same level as the upper ring at this moment and so look as though they are made of that same gooey light. The child's head is a toss up to the system, and the top crest appears to be slightly lit, but the child is also outside the ring. It's shoulders, arms and legs are completely below the lit ring as well as the torso of the adult. All these parts are in the grassy camouflage zone.

Scientists today are experimenting with camouflage which involves projecting the image of one's surroundings or background onto specially designed surfaces on the front of a suit or other surface to be disguised. The effect is much like we see here. An almost shimmery surface which looks incredibly like the background beyond the person or object.

That strange power source or control device on its lower torso, the place on us which we would refer to as our stomach, could also be part of the camouflage gear. That odd spidery structure better seen in my colorized photos appears to have tendrils which trail off in many directions about the body. Possibly it's connected to a type of suit which controls or enhances the camouflage effect. The suit may be so thin and molded to the body that we may not be able to distinguish between the suit and the body itself. All of this magnificent technology creates a whirlwind of subterfuge which blends beautifully into the background noise. It's only abated by what seems his determination to reveal himself in an attempt to respond to our race, which has clearly set out to communicate with extraterrestrials. Well, at least one scientific faction has, and has opened a small crack in the sometimes cruel, aggressive and (possibly) justifiable paranoia of the other factions of the human race towards unknown beings.

Power link or camo-controls perhaps?

He's stepping forward and giving his people just one small hope of communicating their desire to be non aggressive and friendly as well. I'm not a cynic by nature, but I don't give it much of a chance in hell to succeed. Not unless someone with no particular agenda besides telling the human race of their existence comes forward. At least that's what I hear floating on the wind.

And so I ask. Do we really want to throw away this chance? Are humans ready for interaction with another intelligent race? Will we gamble that they are less advanced than we are? Or more....

What if we just continue to ignore them? Will they go away? Not sure.

Extraterrestrials. Strange, yes. And as strange as they are to us, we may also be to them. With our Pioneer and Voyager space probes we reached out to them, hoping they would respond in kind. Here they are. Do we reject them, hate them, chase them off our planet? I say "No! Let them visit in peace!" So bravely, he stands with child on his back. He displays his tools and ornamental objects of status on his belt. No obvious weapons are in his hands. He is not camouflaged completely and neither is his son. For whatever reason, he now stands fully visible to my camera. He waits with my dog who is also patiently posing for the photograph. They seem to know each other already and are at ease. He hopes I'll notice this. He hopes that I can see him now but if not that I may find his image later on. He does not wish to scare me. This is obviously a well planned event. A brave action and an openly friendly one as well. A greeting from one world to another.

Chapter 23

Surprise! Are You Happy to See Us?

The human race has a long history of aggression and conquest. From Genghis Khan to Alexander the Great. From the Spaniards in Mexico to the European settlers and the American Indians. Aggressors with more advanced technology have eventually become rulers. The human urge to compete and conquer is an evolutionary trait which may have helped to put us on top of the planetary food chain. Whereas this is the case on this planet, races from elsewhere may not have these same violent tendencies. But we must assume that they do, or we may wind up like the Aztecs. Dead. Eventually, every race comes across another even more advanced culture. A bit of caution and good intelligence work could go a long way toward preserving our human race as we know it. I saw our interaction as what you might call the parley approach. It was good because a parley has always referred to a meeting where ideas and information are exchanged in good faith. Whereas I may not be able to understand them as I would other humans, in most cases they and we both seemed to able to get our basic ideas across. It seemed like an inventive style of communication, sure, and for the first time we had a friendly interaction. Hopefully both camps welcome peace and cooperation.

Why contact me? Why wouldn't aliens contact some government officials? And if they did which government would they choose? The Americans? The Russians? The Chinese? Maybe they distrust all governments. But why would an alien race distrust political leaders? For the same reasons, I outlined above. History and experience. Political leaders are usually accompanied by and protected by military leaders. The two work together to protect the country as a whole, the status quo or possibly the entire human race. That having been said they would not only wish to meet new extraterrestrial races but to size them up, so as not to be surprised by any aggressive behavior towards us. Unfortunately, if these beings are not perceived to be human then taking them and their ships hostage for study and back engineering their technology might seem acceptable, especially if it were kept secret. If I were an alien, I'd keep my distance too.

What happened at Roswell, New Mexico? No one really knows for sure, and there's no one talking upstairs. Some say that it was just a weather balloon that crashed that day. Later they said it was a super high tech weather balloon named Mogul. They told us that this high tech weather balloon was sent up to monitor Russian experiments with atomic weapons. Some say that an alien spaceship crashed that day. Some say there was one crash site and some say two. Some say that alien beings were aboard, and taken into custody and interrogated. Everyone says that if they were taken, they were never released.

Chapter 24

The Roswell Rat Trap

I have my own theory about the Roswell crash. Alien ships and discs, as well as foo fighters, were sighted during World War II by military sources and kept secret for obvious reasons. Remember that atomic weapons were being developed by both Germany and the United States, and then by Russia after the war. I think we suspected that these discs or UFOs were powered by or attracted to atomic materials and were threatened by man's attempts to build the atomic bomb. They were also very advanced technology.

Then on June 24th, 1947, Kenneth Arnold reported flying discs over American skies. The military woke up again and perked up it's ears. They wanted this technology. They were now on alert. This was their chance.

It is now just two weeks before the reported crash in New Mexico.

· When the discs were sighted again near Roswell, the military put a newly devised plan into action. A specially fitted weather balloon was launched from an unknown site and let soar high into the sky with an important payload. Not just weather equipment this time. It was a new payload designed for just one purpose. They say it was to monitor the use of radioactive materials and it definitely was, but probably not from Russia. It was most likely sent up to sense radioactive emissions and sounds from something so strange and exotic that we were not sure we could actually catch one. An alien flying disc. A UFO. Once sent up, the team would have waited nervously and my guess is that when their devices sensed that a loud vibrating radioactive source was close by, (a UFO), hovering near and curious about this very interesting looking device attached to a balloon, the payload was detonated. A small bomb. The rat trap went snap! The balloon immediately deflated and fell onto the Foster ranch where it was found. Damaged as well, the UFO broke into two pieces, which crashed miles from each other. The balloon and the two pieces of the UFO were later recovered by the military. One piece of the craft held alien life. The balloon, the craft and it's occupants were all spirited away. No evidence remains today, only stories. Maybe that's how it happened. Maybe. I think the military scientists got lucky on one of the first tries. Back then, the UFOs were not expecting any violence towards them. They were safe. They may have also been very wrong.

The military issued a seemingly honest press release stating they had recovered a flying saucer but then later recanted the story and told us it was only a weather balloon. But witnesses saw the wreckage and have come forward to tell us of the strange pieces recovered and of having seen aliens. Captured beings spirited away to military bases? Well I guess that's just the way post war humans treated each other. Secret holding and testing areas? True or false, they certainly were never released. Just how would your family, your people feel about that if you were the one being held? Were there real aliens? Could they have been the same race of aliens as in my photo? The military claims there was nothing there. But if it's true, we sure do know how to teach those darn aliens a lesson! A lesson in human paranoia, aggression, secrecy, and distrust.

What about the Washington UFO event of 1952? Eight discs were sighted over the Capitol and also tracked on radar at the same time. The discs were seen and chased by fighter jets until they moved away at super high speeds and disappeared. If I were the commander of that group of alien ships, I'd be saying something like, Hey! Screw this! Those jet things have weapons mounted on them! We stopped by to see the President, say hello, and what do we get in return? No wave or polite hello. We weren't even offered a piece of fudge cake! Let's go land at that farm over there in New York and try talking to the locals. How about a nice rural highway in New Hampshire? We could just stop that car over there and say hello to Betty Hill. Now that's got to be a lot safer than this merry-go-round! I bet they won't even send those jet fighter things after us! We could even head south and stop in Gulf Breeze, Florida and do some diving, maybe take on some supplies? Try the grouper! All kidding aside, I'm not certain anyone, even an advanced race would want to continue to go where they're not wanted. How to overcome this prejudice? Make contact with average people and attempt to be friendly, non-aggressive and non-violent. Try to build up a small following. A group who knows and can testify that you're kind of..... alright in your weird alien sort of way.

And just how does a race of beings so different from us that they possibly do not even use their mouths the same way we do accomplish this feat? We do not speak the same language and possibly don't even produce sound the same way. I mean we are really different, and many people even report something strange like telepathic speech, and this could be the way aliens have evolved to communicate. Of all the animals on our planet, how many can you think of that have developed a sophisticated verbal language?

One, that we know of, so far.

Besides developing a large brain and walking upright to free the use of our hands to make tools, our jaw bones, teeth, the muscles in and around our mouths, the tongue and our larynx as well as our breathing must coordinate to produce complex speech. Our advanced brain must be intricately involved as well as our ears, nose and throat. No other animal on the planet has all the necessary elements developed and put together to form that one perfect package. At least as far as we know. So the chances of another race being able to produce the same sounds as us are slim at best. But since UFO's and their inhabitants have been reported throughout the ages, they have had plenty of time to learn and understand a variety of languages. And so that makes telepathic language one step easier if telepathic communication depends upon knowing your contacts language. The universal translator would come in real handy right now and maybe they have one. So however this works, it seems that the aliens at least who fly around in the saucers, seem to know how to communicate with us and are giving us the ability to communicate with them to some extent. Now that in itself is cooperation! The rest of most peoples interactions are not quite so fun, but the majority have not been intentionally harmed. If you believe them at all, experiences range from trips to other stars to medical examinations. Then the aliens just let them go. So why medical examinations? This seems to be a more recent development within the last few decades or has at least increased in alien popularity during that time. Why? If we're going to be friends, they must be certain we aren't carrying something which will kill them. First of all wouldn't we give an alien a physical exam to ensure they were not contagious and also to understand how all their parts worked together? If they'd let us do that, we most likely would. A vast amount of knowledge could be obtained from such an exam, but of course there's the old saying that human scientists will not believe in anything that they can not put on their examination *and* dissecting tables. That would scare even me! And as for the Roswell aliens, it's probably the reason why we never let them go afterward. Pieces don't walk well. So much for blind trust.

I suppose I should get used to the fact that many will dismiss all this instantly without even reading further. Oh well, ho hum. At least you're here.

I should think that there would be additional interest in humans these days due to the huge advances we are making in every scientific field. Just maybe they are feeling a bit of alien anxiety in regards to this new level of technical expertise we are exhibiting? From an upright ape with an unimpressive brain size to modern man in just a few thousand years is mind boggling. It took us millions of years to begin walking upright and then in a few thousand years here we are, large brains, sophisticated tools, skyscrapers, computers, nuclear weapons, and quantum mechanics. The point is that the aliens who I believe have been stopping by to say, "Hi!" over the last few thousand years may now be much more interested in us today than ever before and may also be somewhat alarmed. We could now present a threat to them. Better to get on our good side as early as possible. Best to get on their good side as soon as possible.

Chapter 25

Solid Or Not,
Gentle Aliens! Start--Your--Engines!

The saucer from Roswell was reportedly made of solid material. Many crash parts were collected and subsequently spirited away by the military. But many UFOs are said to control the ability to slide in and out of our physical world in part or in whole and so do not seem to be very solid at all. Why would this be?

For an alien ship, being made of completely solid matter would not be very wise on this planet, where men with large guns, high power lasers and surface to air missiles might just set their sights on you. In ancient times, it was only bows and arrows, rocks and spears you needed to watch out for, but these days? Look out. For a non-violent race, acceleration and speed would most likely be the technology of choice to avoid all these things, but the aliens would need to stay ahead of the game. The human race is making progress quickly, to say the least. If aliens are trying to make a good impression on us, (a newly technologically conscious race) high speed alone would allow them to evade our weapons without taking aggressive action towards us in response or in defense. Many UFOs are also reported making such impossible maneuvers as right angle turns and to accelerate at speeds which would kill a human being instantly. They can accelerate so quickly as to be out of sight in the blink of an eye. These impossible maneuvers are compounded by the ability of some ships to become invisible upon demand, or when certain that they've been seen for a bit too long. They may even be more than invisible. They may be able to slip out of our dimensional space and just be "not there".

We don't have a clue as to how they are doing this stuff, it's off the charts. Well out in front of our scientific knowledge and will be for some time. I'll be sure to let you know when that changes. That of course is assuming that scientists regularly and routinely update me on all improvements and inventions in the physics and military worlds. Ha! I'm lucky to pick up the right magazine or hit that lucky internet site with some hint of technological advancements and extrapolate the gobble-t-gook in my own feeble brain to allow me to see into the minds of our scientists and military leaders. Right! What are we doing right now in this area? What kind of advanced weaponry and transportation do we have in the experimental or just plain secret realms? I don't have a clue, not having had my top secret clearance updated this year! I sent in the form, and a check for $9.95 but whether lost in the mail or whatever, it has not arrived and neither has my decoder ring! So I have to go with what I've already got in my brain for today, which isn't much and is thoroughly amazed by what these little alien guys can do with a few odd molecules, an energy bar and some spare time. I'll say one thing just one more time. Their technology of choice seems to be speed, acceleration, and camouflage, not weapons. Weapons, they seem to be telling us, are in this case, at least with us, unnecessary and possibly even childish.

By the number of sightings of UFOs, I would say that they are also definitely watching us closely. Either that or they really don't give a rats ass about us to begin with and are just hovering about where we sometimes catch a glimpse of them doing their alien laundry or something. We may be just a curious life form inhabiting that pretty blue planet round about, and they think that well, someday we might just evolve into something useful. Hey, maybe they just decided to say hello to us for a laugh! We keep sending our probe things into "their" space, so they come round and swirl up our wheat fields! So assume for a moment that they are watching us closely. Why? Possibly just because they..... can.

How do they do this? Is it all extremely advanced technology? Let's suppose that this extraterrestrial race which I interacted with, is made of a very strange stuff? Possibly they're not the same guys in the solid ships at all. Just suppose that they are actually made up of a material that does not normally interact with our matter well or interact with at it all under normal circumstances. Did I just say "normal" circumstances? Wow, what's normal about anything in this book? Not much. And what is normal? What is abnormal for that matter? And for that matter is solid matter normal? Does it matter? To your average humans, it certainly is normal for matter to be solid and it certainly does matter. But matter is far from solid and is actually, mostly empty space. Your "average" human doesn't think about this because when sitting down in a chair for supper, the last thing you want to think about is sliding through and landing firmly on your ass on the dining room floor. Assuming the floor is more solid than the chair that is. People need to be able to get about their daily lives without worrying about such things as why we just don't fall through the boundaries of the materials we come in contact with? The boundaries, yes, the surface, the edges, the seat of that chair you just sat upon. We don't even think about it, we just don't fall through. And furthermore we know exactly which things we can sort of fall into such as water and air, and we keep these in a separate place in our minds. These things are obvious to us. We grew up since childbirth learning which ones were which. We also learned that we can usually see through air and water and light illuminates them and anything within. Ooooh. We also know that we can hear "through" air and water, and that sound helps to locate objects within them. This is how we understand our space. Solid and not so solid. Now I've neglected a whole bunch of other items of interest but for the sake of expediency I did. So if you take all this and put it together, you see that solid objects are pretty calm, and we expect that other solid things don't move through them. Air and water are not so calm, and as objects move through them in our reality we have ways to discover what they do. It's called using our senses. We watch for anything moving in our spacey little world by catching light reflecting off of it or by listening for sound coming from it. Since we expect everything to behave according to these rules and anything worth paying attention to will also be a blaring alarm to our senses, we then tend to create another sublevel: Those items too insignificant, nonthreatening or way too stealthy to capture our attention. Most things fall into this quiet realm.

Into the background noise.

What if these UFOs were made of something which we see only as background noise? Consciously it just doesn't register in our minds. Wouldn't that be a great way to watch us inconspicuously so as not to set off everyone's alarm? Could they be, compared to us, closer to the density of empty space or have less solid boundaries? Could they exist in a skewed set of dimensions where we can not go without help but from where they can see us? Could they be made of that mysterious stuff called dark matter?

Dark matter does not interact with our normal matter, nor does it interact with light. It does however, respond to the force of gravity. Maybe this is how these aliens are able to blend in so well without being seen and also seem to be able to sometimes move through solid walls. At least in an orb they can. Standing next to Greggor, the alien seems to be able to keep his balance just fine. His orientation to the ground would lead one to believe that he is standing firmly on the grass and is having no trouble staying there whatsoever. Gravity does seem to be acting on him just as it does on everything else in our universe. Yet gravity does not keep his ship on the ground, but neither does it keep our jets there either. Their alien craft's acceleration and right angle turning could be problematic for this idea, but that could be overcome technologically. Superior alien technology. There are definitely weird light effects, and when necessary light does not seem to reflect or refract from their surface, making them invisible. Also, under certain circumstances, which they have learned to control, they may also be able to communicate with us and interact briefly with our normal matter on this planet. The communication probably takes the form of subconscious thoughts or "self-talk" or perhaps occurs in dream time where we routinely push it down to subconscious levels immediately upon awakening. Possibly they can pass into and out of our "reality" or our set of dimensions at will becoming a bit more solid or obvious at one moment and non-solid the next. Possibly, they are made of a higher concentration of energy and less matter than we are. Energy beings? If they have the ability to control their density, energy level and dimensional orientation, it would explain many of the sightings and experiences of witnesses over the years. In WWII, American fighter pilots were confronted by what they called foo fighters. Glowing orbs of light which may have been annoying but were none the less rather harmless to our planes. In one instance, a pilot reported that they actually flew straight through one of these orbs as if it did not exist. Now, possibly this is the aliens advanced technology making itself known, but if so, why be so non-interactive? If they are so inconceivably advanced, then they could just as easily blow our planes right out of the sky without even blinking an extraterrestrial eye! They could easily maneuver out of the way of our planes instantaneously. Why just let us see them and then as many UFO's are said to do, turn transparent and disappear? They obviously don't want to fight or have no weapons. Just maybe they are our friends and maybe they want us to know that they will not intentionally harm us. Possibly our fighting, warfare, and development of advanced particle weapons is troubling to them.

UFO's are reported in all shapes and sizes. Many are saucer shaped, but others may be tubes, balls, orbs cubes, triangles or ovals. And so, I am not saying that all UFOs are made of something close to empty space. In fact, the reports which I have read, seem to indicate that there may be many different types of craft with differing abilities and technological tricks. This would roughly correspond to the differences between a luxury sedan and a tank, a motorcycle and a garbage truck, a jet plane and a kite. Some of these would seem to be perfectly solid, and land allowing their travelers to exit, walk about and sometimes commune with the locals. Some do not. To simplify things, I am only going to look into the beings which seem to like the orbs, the ones which I have personally encountered. Why they like orbs, I do not know. Balls of light are really cool I guess, and oval orbs are just to die for. Maybe they're the sporty model. I suspect that these orb entities are all from the same basic place whether it's an alien world from another solar system or from our planet using their own special set of technological inventions to alter the way our dimensional space interacts with theirs. Maybe they're from a big fiery planet some call " Sol ". They may also be from a parallel universe which has slightly different characteristics from ours. Possibly matter is not as solid to them as it is to us.

I know there are amber-white orbs of varying sizes. Some would hold a full size being such as the adult in the photograph and others seem to be too small. Many of the glowing orbs look like they are made of the same substance or material. They glow with an internal light and are often amber-white or sometimes bluish white in color. This would include the so-called foo fighters, balls of light over crop circles, and lights encountered which do not seem to be attached to a structured craft. Why would there be so many different varying sizes of the orbs? Let's take the earthly example once again. The three to six foot orbs would be the perfect size to fit one alien into, such as the large one in my photo. The next size up could be for multiple aliens and the massive size for group transport. The "bus" so to speak.

The fact that the larger the orb, the higher it seems to fly would tend to support this thought. The only issue I have with all this occurs when I look at the smallest of the orbs. I have personally seen one that was approximately golf ball size! And as reported with the huge craft, it had a texture of sorts, similar to a walnut. I got a good look at it's globular surface as I watched it hover, move horizontally and vertically and disappear into my wall. This orb did not appear to be afraid of me but displayed a curious nature, almost playful, and stayed until I got up and reached out for it. It then moved away, flew right through that wall and did not return. Now, not being really good with rejection, it bothered me the rest of the week. I really wanted to interact with this thing and was being very nice to it, but unfortunately it did not understand this or was under instructions not to talk to unknown humans and left. I felt that this was blatant extraterrestrial discrimination and I was rightly pissed off, just after getting over being sad that I'd been dumped! That was the point when I remembered the little alien in the photograph. And the thought suddenly hit me. Could this orb carry a tiny alien child? And if I were an alien parent, would I allow my little aleikid to hang around with a strange human in his bedroom. What might he see? What might happen? So I figure that Mom and Dad probably didn't know where the little guy was and not knowing whether I would tell, it left.

So it's possible that the smaller orbs could actually be carrying small aliens! We know now that they do have their children here so small orbs might just be the way the little buggers get around. The small orbs do sometimes seem playful. Could it be that this is their method of letting their kids go out and explore? "Oh No!", you say "They might get lost!" But what if the alien parents had a bit of parental control technology built into the craft? It might be that the kid can fly around in it but can not land, or allow anyone else to touch it, and it will auto pilot back home within three hours, just before milk and cookie nap time. Just press #3 on the remote! *Zeeep!* Other than that, the kid can fly around free. Could small orbs be the alien equivalent of a baby bouncer? I certainly do not know. We'll have to wait and see!

Someday we may find out. I think back to the reports of orbs visiting or participating in the formation of crop circles. It might just make sense, for just like human children, the aliekids may be very interested in creating designs, shapes and patterns. Our fields may be just one big alien art board, and we are the lucky ones to be able to see their works, admire their intelligence and to investigate and imitate their designs.

Far fetched you say? Think again! Look at the photo! These are living creatures, with lives and kids and creativity. Where they come from, may not just be the sterile alien world we imagine where everything has a purpose that's scientific and military. It may **not** be the cold, dark and wet domain of monsters. They just might be more like us than we think! Seemingly without the aggressive military thought process.

Another idea is that for whatever reason the orbs may be able to change size. There could be some advanced technological way that they could collapse the space between molecules or utilize their knowledge of space-time dimensional goop, to just transform their craft or whatever you might call them into mini-orbs! We're talking something that sounds like science fiction here, but hey, there's glowing aliens in my backyard, OK! And they're not being real solid, either! So go figure and keep your mind open. Possibly it's not technology and is just the matter of fact way that they travel around, and size is something that's only stable and permanent for us. This again would be closer to magic than what we feel comfortable with. So if advanced technology could appear to us as magic and beings inhabiting a skewed set of dimensions different from ours and doing what they do appears to us as magic, we have a problem. In other words, don't take a whole lot of time trying to figure it out because it looks like it's gonna take a very, very long time and you are not going be able to figure it out.

Maybe one of these guys will eventually tell us what's going on. But, don't hold your breath, I've already tried it. They don't seem really open to doing that. If I asked you to tell me how to build a jet airliner like the one you just flew in on from New York, or even your new car, could you do it? No! They probably just exist with their technology having read the "quick start " guide to orb operation and haven't the foggiest idea of how it's constructed. Some engineer at a manufacturing plant is laughing right now. So why do the craft change size? Maybe just to keep us amused and wondering how deep their wormhole goes. Being alien, it would be easy to mess with the humans. In point, every once in a while you'll get one of them playing practical jokes on us and doing something like proposing we start a new religion based upon some old sushi plates with scratches on them. But frankly I'd rather eat off the dam things, and when both of us have a nice full stomach and the kid is out playing, have them tell me what's really going on. Right up front so that any idiot human like me could understand it. Teach me a little magic please and maybe I'll let your guys stay in the guest room.

Personally I am getting the impression that the orbs simply change size at just the push of an alien button. Or more likely the initiation of an alien thought. Who needs buttons in your alien magicatechno world? I say this because the orbs, assuming that they are actually craft, sometimes appear to change size as observers watch! Yes! The orbs seem to grow or shrink to a much larger or smaller size very quickly. Almost in the blink of an eye. So, let's say a four foot tall adult alien creates an orb to travel in. It might be six foot round but could also shrink down in size and be able to appear to us as just one foot in diameter. If the kid in the photo is in an orb, it may only appear to be a few inches round. Hence I think I was visited by junior! And at play. So the next question is, how confident of your technology and such must you be to allow your kids access to another world inhabited by huge numbers of very large humans. Very confident! We must accept the fact that these aliens have no fear of us whatsoever! Furthermore if their technology or abilities are as advanced as they appear to be, we need to be at least careful with these interactions. I was doing my best to be "friendly" and attempting to glean any magic tips I possibly could. Unfortunately, verbal or telepathic communication was not something which was easily accomplished to say the least. It was torturous! I do not recommend it to anyone.

What would the limitations of their size changing abilities be? Who knows? Could it be possible for them to scale down to the size of a pea or a pin head or a molecule of hydrogen? I was sitting on the couch one night, and to my surprise I suddenly heard a solid pure tone in my left ear. The television was turned off, and there were no other loud sounds in the room. My friend made no move, so he apparently did not hear the tone, and as I poked my finger into my ear it occurred to me that it possibly could be an alien ringtone going off in my head! Mom calling junior to see that he'll be home for dinner? The horn of a tour bus heading off into the gray matter of my brain to see if there is room to store a bit more math or something ridiculous like that? Maybe the little guys were just playing around with my nerve endings and shorted me out for a second. Whatever.

So anyway, as I stood there with my finger poking into my ear, I suddenly realized just how silly I looked. And as my friend tilted his head sideways and asked sheepishly, " Are you alright?" I snapped out of my stare and quickly withdrew my finger. As I struggled for an explanation, I also thought that what a great idea for the medical community it would be, to have miniature live alien physicians on staff to shrink down, go in and zap stray papillomas, prions and previously impervious predatory parasites! Give em a ray gun and put em to work. Fantasy? Sounds like an old, bad sci-fi movie doesn't it? Maybe you could hire them to watch for, chase and use their ray guns to disarm nuclear missiles fired from uncooperative, rouge Asian nations. How would we pay them? What would they want in return? No way to tell. Great idea huh?   Oh but there is absolutely no evidence that alien UFOs can shrink down to those sizes. At least none that I have personally seen to date. As for them entering and rummaging around inside the human body, I can only say that I am of sound mind and have no miniature alien beings in my brain right now. I think.

PART III

THOUGHTS
AND WORDS

Chapter 26

Alien Communication - Telepathy?

Torturous is the word. Yes, I do believe that would be the correct word! But I have heard reports of interactions between humans and aliens where the communication actually seemed quite easy. If you believe them, Betty and Barney Hill did not appear to have much of an issue with it at all and had some great conversations whether verbal or not with alien beings back in 1961. The Hills were abducted while driving on a country road in New Hampshire. Aliens supposedly performed medical examinations on the couple and communicated where they had come from utilizing a star map.

If it were being scored, sounds to me like both parties earned an "A" on the use of their alien interactional skill sets. Not an A+ but a great grade nonetheless. But it could be that not all aliens are alike in their language skills, education levels or universal telepathic translator technology knowledge. Maybe I ran into one who just didn't take time to read the translator instruction manuals thinking that he would never need to use the damn contraption anyway! Maybe he simply flunked out of Extraterrestrial High. At its best, the communication from them was like unknown irritating noise and at it's worst it gave me one hell of a migraine for days. I felt as though a crazed clown or criminal disc jockey had invaded my brain. Hold on, it could just be the neighbors playing their rave music again! Oh, and then at some point the child began to communicate, as well. What a mental mess! Whatever it was they wanted to tell me often came through as a meaningless collage of bright sounds and loud images. For quite some time, I found it really tough to listen to, but I realized that this was a new experience for me and maybe for them as well, and I just had to keep going. It was way, way too interesting. Way too cool. Who knows, maybe I am the first to actually try to get through teaching an alien to communicate with a human and for them to teach me to think back or speak back to them from scratch. And I do mean from scratch. One of the things which I seemed to understand through weird broken imagery from them is that the larger, more complex "ships" have the ability to "help" them communicate much more easily. Maybe there's a device which senses alien (in this case human) presence and probes their brain waves ascertaining language structure automatically and then does the math to bridge the communication gap. Or they may just have someone there who is schooled in communication, so everything goes along much easier. Perhaps they've had time to put this alien techy guy through human speech 101. We have been transmitting all sorts of stuff into space for decades. Maybe even some home course on language skills? Anyway, bridging the language gap seems to have been a bit more difficult for this team than for others. Great! Why me? Why couldn't I just get the "A" team? I'm a decently pleasant guy, willing to try just about anything once and learning a new language is not too far out especially if you figure you're possibly the first to try it. Plus humans have been able to create and speak or at least read any number of cool languages. Hey, we've pretty much mastered just about everything from hieroglyphics to Hopi! But why me? I just want to be able to say "Hello!" "Good morning." And "Could you please direct me to the bibliothèque?" properly, for the most part, anyway. I suppose if their machine didn't quite get the pronunciation or subtle accents right I could survive, and they would probably understand me anyway. That would have been nice, and I think I deserve it too. But nooo, I have to start trying to learn from scratch with no assistance and no universal translator machine-thingy to help. So I tell myself I am going to do this even if it takes me most of the rest of my life which at this point I am hoping will be at least a few more years and I won't end up in some alien stew or on the dinner table for alien holiday feast. That is if they even have an alien holiday feast. I felt like a turkey. I also later on grew to accept the fact that they DO NOT EAT HUMANS. Phew! That was close! So I did my best to understand and learn, and I figured I'd take a little kidding and laughter along the way. Oh you talk like a human! Like you've got too many teeth in your mouth. Anyway, they seem to understand human English at least a bit now which sure comes in handy. I, on the other hand, never did learn the least little bit of their language. The longer I spent working on it, the more difficult it turned out to be. So then I figured that they are the ones who came here. I'm the host. They should speak *my* language, right? Why yes! Yes! So, listening to them use "sort of human" and responding in human turned out to be the interaction of choice and by far the most efficient for a while.

But most effective is far from exact and quite far from pleasant.

Eventually, it did get a little easier.

Because eventually we gave up. Spoken communication? It is not easy to speak or understand another being's language if you do not have the same sound generating (mouth) parts. A cat can understand many words humans use and yet as hard as they may try, they can not repeat them. Those fangy teeth just get in the way, and the lack of good lips is an issue. Maybe they could work past it. Maybe we should learn cat. That's another story.

Eventually, we formed something like a teledreamapathic way of circumventing the whole thing and that's pretty much where we are today, working in multiple dimensions. And it works pretty damn well if I do say so myself.

Could I be imagining the whole thing? Were the attempted conversations all in my head to begin with? To my surprise, I've searched my mind many times only to find no firm answers. I pride myself on being routinely skeptical. That having been said, I am also not so bullheaded that I sway towards wanting to believe that every unusual event is actually rooted in normally accepted, and usually outdated everyday 8th grade science, so I believe they were real.

How do you know if someone is communicating with you telepathically? I'll tell you. In a minute.

Telepathy. It's not an easy thing to understand, and it's near impossible to convince someone else that it's actually happening to you to boot. The substance of it for me? It's thoughts. You have what appear to be your own thoughts. Pretty easy huh? No. Not really. Because, these telepathic images, words and sounds can cascade through your mind, more like leaves carried over a waterfall. They can take the form of images from your past or sometimes what you might think are imaginings of the future. They can take the form of pictures like photos in an album which float in sequence over your thoughts. Those images are more like words that combine to form loose sentences, which grate at something meaningful for you to chase after. Those sentences can either accent, describe or defeat your own train of thought from one moment to another, and I found them to be more like an old roadblock on train tracks than anything else at first. The old roadblock sits on the rails demanding that you stop your forward movement. Stop your train of thought. The wooden boards creak and groan. The metal nails have just about rusted through and press in and out of their holes with the fleeting stresses of the winds upon the wood. Stop! Go no further on this path! The words are so solid, yet misleading, they attempt to tie down your immediate thoughts. For a moment, your mind feels like an insect trapped in amber. The words adhere to the rough wood of the roadblock, but as the wind swirls about they are eventually torn to pieces. Only the thoughts and images float over the top scattered about by the wind.

The roadblock itself shutters in the wind and small pieces fly off. A board eventually breaks and falls to the ground leaving a small opening for the wind to pass through. It carries more info but not enough to grasp the exact meaning. Eventually, the wind streaming through the holes, rattles and shakes the structure to the point that it simply falls apart allowing the main telepathic communication to flow past. The images begin to swirl into a whirlwind which feels now almost like human language. And as each moment passes, more and more images and a few words form until you finally get some partially clear meaning out of the gobble-t-gook dripping from your gray matter. Crystal clear goo. Make sense? This all seems to happen in the blink of an eye at first and then carries on as long as the communicator sends. Only sometimes does it get through clearly.

What it feels like in reverse, heaven only knows, considering my failure at alien-speak, I'm just glad that I'm mostly on the receiving end. Which would tend to make you believe that either they are more intelligent than we are, or they have a brain better suited to telepathy. They may also have some tech assistance which they aren't telling me about.

I'm calling this communication telepathy for now because I really don't have any other explanation for it and the word "telepathy" has a great ring to it. It fits into every conversation concerning aliens whether they're invisible aliens from Mars or just your run of the mill extraterrestrials from some other dimension. What else could it be? It sounds good. It works. I'll use it till something better comes along.

How do you know if someone is communicating with you telepathically? Again, all I know and all I can tell you is what I experienced. But I'll do my best to put forth an answer.

Here we go.

First of all, your thoughts, which normally would be of common everyday events in your life, start to transform into something which looks the same on the surface, but feels like it has a slightly skewed context. As you continue to think your everyday thoughts of simple tasks and surroundings, other words and images break into and float on top of your own thoughts. It's as if you are trying to focus on something in your mind, but someone else is pushing a different play button on the remote. You suddenly have two distinct and different thoughts playing in your head at the same time. Very disconcerting to say the least. The tough part is following your own thought process while simultaneously hearing another going off in a separate direction! I found that when it happened I usually found it easier to give up on my own train of thought than force myself to ignore the other input and continue on my own path. But I made a game of it and after sometimes long and stressful mental encounters, actually did make a bit of headway towards holding onto my own thought stream.

Everyone thinks to themselves. It's the same thing as reading to yourself. You hear the words in your head, and there's no need to use your vocal cords or your lips. We all know how to do this. It's a human thing. Many people can read a book and listen to a movie on television at the same time. It can be done, but I find that most people have trouble focusing on more than one thing at a time. The television plays in the background, and if something interesting happens your subconscious will alert you to look there instead of at your book. But it's just too difficult to actually read your book and also follow the plot of the movie at the same time. So here's another example. As you are reading this book, reach over and pick up the magazine laying beside you. Set the magazine down just in front of the book and attempt to read the magazine at the same time you are reading this line. Can't do it? You're not alone. Your brain and eyes want to focus on only one train of thought at a time. They recite only one line of words in one sentence at one time. Now hand that magazine to your friend. Hopefully you have one nearby. As you read this paragraph, have your friend pick a page and any paragraph and begin to read aloud. Attempt to concentrate on both and hear both paragraphs at once. One playing in your mind through your eyes to your brain and the other playing there as well through your ears to your brain. You are getting closer to understanding. Some people can do this little mental multitasking trick for a few seconds and some can not.

At this point, we are only dealing with human words and of course we are also only dealing with physical, accepted delivery systems. Namely our senses of hearing and seeing.

Now imagine that some of the content of the paragraph your friend is reading to you is not words, but merely a photograph. Wow, what did he just say? Your friend is projecting images of the content of the words and images on the page. The photograph is like a hieroglyph in that it may be a whole thought or may be just a piece of a whole thought. If taken out of context, the meaning changes significantly. So if you break contact you may misunderstand or not fully understand the message presented to you. Now you have your own thoughts taking place and some words with images layering on top. The words and images don't look like images in a dream because they are waking images, similar to those remembrances of a place where you used to live or someone you used to know. They are fleeting and not quite solid. They start to form more words. You must refresh them over and over again so you can see them clearly, but sometimes this just isn't possible. Since they are not your thoughts, your "friend" attempts to do this for you. Their intention is to ensure you get the entire correct message.

In this example, the interaction is human, but in my experiences the interface and cooperation between minds is alien. So alien that the gentle repetitive helping hand becomes a turbulent overactive instant replay of alien imagery delivery. A fibrillating muscle clamped about my mind.

No fun.

But, I never once felt as though I was being attacked in any way. My only conclusion was that we were both in a struggle to communicate. So how do you know? First of all..... you just do. It's totally different from listening to your own thoughts which you are creating within your own mind. Words layer on top of one another at times. Since another being is sending to you and thus involving your mind, they do not have your private access to your own internal voice. The over layered voice usually sounds like a memory of someone you once knew. Your brain seems to have a choice of who's voice to use since yours is currently occupied. Or they do, because of how advanced they are. Am I crazy? I suppose I could have brain cancer in which case I should get to the hospital soon because it's obviously already out of control. Or my right brain is acting up and challenging my left brain to a game of mental ping pong. Or my left brain being overly concerned with language is trying to out speak itself and amaze me with its ability to carry on two diverse conversations at once within my own mind. The CIA could actually be feeding me LSD, and I might come down soon. If it is the CIA, I truly hope it's been a very long wonderful trip. I hear they make good drugs. If I wake up soon, I really hope there's someone next to me when I do.

At the tone, please choose another random voice. *Zeeep!*

Does all this seem a bit odd to you? A bit alien? A bit like a dream? Well, you're not alone, and I must apologize. I realize that I am sounding a bit crazy, silly, scary, comic, terrifying, mystifying, sometimes informative and just plain weird all at the same time. Those persons who have dropped out are obviously not capable of comprehending the complexities of alien contact. You, on the other hand have been here all along, and I assume you are remembering where you were from day to day even if you don't use a bookmark. If you do, do you remember the previous chapters? You do? Good! Bingo! You get it, and it sounds like your brain is handling this just fine! That's alien 101. Congratulations! You just graduated! Throw your cap in the air! Yeah! But, tell me. Where are we? Why are we all here? Why am I still sitting here naked? Why don't I just go and make myself a nice hot cup of tea? I do believe I've been influenced by Monty Python..just a wee bit. Thank you M.P. You've kept me sane!

The whole thing is very much beyond belief to me. One day I invented a little mind game to help me try to understand and establish whether this reality was at all possible. I imagined that I had never seen any other intelligent creatures before, and I was out one day, passing time, just looking around for a new intelligent being to discover. (Or for them to discover me). I happened upon an ocean. Haven't you ever happened upon an ocean before? The cool waves were pounding the shore and seabirds flocked to pick at insects on the wet beach. I boarded a small boat and headed out into the open water. After some time, I came upon a group of beings swimming in the water. Ah ha! One of the beings came swimming up to me. I knew immediately that it wished to communicate. I called out that I would like to meet it and would it please come over so we could talk. I gestured to it with a wave of my hand. I had seen it gathering food, and I had some on board, so I picked it up and offered it to the being, as a gesture of good faith. The being swam over, and sitting up and smiling, it seemed, said, "Hello," I suspect in it's own language and took the food from me. It immediately swam in a circle and spoke another word or two, then jumped into the air and threw itself into a somersault. Upon landing and a bit more swimming it picked up a small piece of seaweed and threw it to me. It landed in the boat, and I uttered "Thank you" in response. I wasn't sure if it knew what humans ate, but I suspected it didn't. Nonetheless, I smiled and felt quite pleased. It spoke again. Now, I'm thinking my language is of human words, but his is of high pitched squawks and clicks. I waved my hand in a gesture of peace and greetings, and he sat up in the water and also waved his fin-thing. I do not know whether he could ever possibly understand my human language or that I could ever understand his series of circular movements, clicks and whistles. But communicate we did. We both knew for a time exactly what the other was saying. A primitive little game, yes, but it opened my mind to letting the process work, though most times seemingly disorganized, unfocused and sometimes illogical.

The fact that our languages were completely alien to one another did not prevent me from noticing that his was also verbal, and nuances were conveyed through body language, expressions and sometimes actions. What seemed to me to be a few squawks, clicks and squeals, could very easily contain a multitude of variations capable of carrying meaning comparable to human language. My feeble utterances, certainly seemed just about as baffling to him and most likely made almost no sense whatsoever. But he would also be able to imagine some unknown structure hidden within, as well. Now we have an understanding, but only a faint glimmer of an understanding that we probably both speak a rich language, but have no basis of comparison to bridge the gap towards understanding each other beyond conveyances of basic needs such as to verbally greet each other, to share food or exchange of gifts.

Upon further study, I find he has a huge and varied collection of sounds and over time he also becomes aware that I do as well. The tricky part is that he and I live in totally different environments, so common references are few, and we are both completely unaware of how the others language actually works. What if it is only partially based upon sound waves produced in our bodies. Could his be based upon other things like the sound of a tail fin moving in a particular pattern in the water or a sound and a movement combined, or a telepathic image combined with a head movement? The combinations are endless. We have different muscles and bones to produce sounds and also different brains to interpret those sounds. Do we even hear sounds the same way? Without knowing my friends anatomy, I could be blind to some form of sensory science unknown to humans. The sounds he creates could actually be more like vision than language. How to bridge the gap? Is it possible?

If you extrapolate this to my real alien visitors, the same observations are true. And once again, is it truly possible to bridge the gap?

My conclusion. It must be. We just need a Rosetta Stone, but who the hell was in charge of that little bit of planning? Eh? Where is the big ship, the "A" team, anyway? Probably off somewhere getting a bite to eat, watching the silly humans hanging upside down, screaming their fool heads off at the State Fair. As the ride goes round and round, the aliens shake their heads and munch on deep fried Twinkies on a stick.

The "A" team. Wherever they are, they're not here.

I am not going to get into this too deeply, but I have not only had some interesting interaction with them but also began to understand that their irritating quirks were unavoidable and so I needed to develop my own protection device to keep my mind safe from unwanted incursion. No..... not a foil hat! Come on now. Be real. I am not going to walk about wearing a foil hat unless it's designed well enough to keep out unknown alien transmissions, and also looks really cool. You sometimes see paper hats that look like boats. The origami type precision folded works of simple art that kids sometimes wear in kindergarten. That's exactly the type I won't wear. Oh, unless there's a really good reason to. I suppose that if aliens from Planet X invaded our world and began sending out messages designed to subjugate the human race, and the only way to block the damn buzzing was to fold up a foil hat like I learned in my first year of kindergarten, I would most likely do it. Now a foil bowler or short top hat might look acceptable. Depends upon just who folded it. My own uniquely devised protective device is verbal, language based and strong. And believe me you need it when for a while all you hear in your head is weird stuff repeating over and over again.

The initial telepathic message was something like this.

YOU WE ALIEN DO LAND SEE GOOD STOP CHECK FRIEND NEW PLEASE OFF!

Maybe their A-team was at the park or something that day as well. They returned full of green tea ice cream and looking clueless. You see there are times when the "foil hat" would come in pretty handy. I actually prefer the "cone head" style to the folded hat, but it does look more like someone wrapped you up as a turkey to put in the oven than a protective human head covering. Maybe I'll invent a good looking functional model myself and get ahead of the next alien arrival a bit.

Hold on. They have no plans to invade. I'm not kidding. NO INVASION.

And it doesn't seem to me like they have any future plans to invade either. I've said it before. They just want to check us out, make planetary friends and at this point are most likely happy as hell just to be here alive.

The problem with the telepathy thing is the tortuous constant repetitive irritating talking inside ones head. At least it was for me. Try having someone repeat that bold typed sentence over and over continuously about a thousand times and you'll see what I mean. They tell me they need to maintain contact, or my primitive human brain will just give up and find something better to do forgetting all about them. Since I'm really trying to cooperate and a major fugue event is not on my happy list right now, I go along with them on this.

What was I saying?

Oh, You must remember that I'm not just dealing with a calm adult alien being who is attempting to communicate simple but important information to me. There's this child there as well. I have some friends with small children at home, and after visiting for some time I decided that I would rather strangle myself with my own entrails than have to listen to the little monsters for any length of time.

The adult, who by now I've nicknamed "Sparky" is really pretty much accustomed to speaking/thinking/transmitting to a human. OK, Go ahead and laugh but "Sparky" just seemed to fit. And Sparky has now been at my home for some time and has become part of the family. An invisible part of the family but part of the family just the same. Why invisible? Camouflaged? Probably because they're afraid that someone would see them through a window and have heart failure. They really are very considerate that way. It's not every day the new neighbors move in, and they're aliens. As far as having them at the house, I was, at the time, living alone, so the addition of a couple of aliens was not a strain. It was still a very small family. I know what you're thinking. Let's keep this clean. I'm not exactly sure where they slept, but there has been no sex with alien beings and no rectal probing that I'm aware of. Nor have they covered me in green goo and replaced my brain with an alien hot pepper plant before having me write this book against my will to help implement their covert alien agenda. That would be to have all humans send $24.95 to Save-an-Alien6 Foundation to help support the poor wretches who crash landed at Roswell and cannot find real work. Starving and oppressed by our human government they toil in slave labor camps constructing foil hats. Enough kidding around.

No. It's all really tame stuff. Weird but tame.

Let me back up a bit.

Once I had discovered the images in the photograph, I began to connect all the weird stuff which went on just before and after, and then began to notice the strange images and weird thoughts in my head. I was very busy with almost no time to concentrate on this stuff, and one day I just heard him clearly. Now it still didn't make any sense to me as far as human language but there was something else besides me swimming around in my head. As I examined this phenomenon, over time I became frustrated with the fact that I could not commit enough of my time and thought process to it, to really make a difference or to make it make sense. But I was convinced that it was not just my thoughts and fairly certain that I was not being dosed with LSD.

I told a couple of close friends about the photo and some of the strange stuff going on, but almost to a person, no one believed me. I finally had told nearly a dozen or more people, and some refused to even talk to me about it days later. One just laughed each time I mentioned it but would not comment further. I suspect they all thought I was either nuts, or I was playing a prank on them. This is where we have come to in our society. We reject reality and favor false illusion.

OK. Make believe you are me. So, you make the decision. You have proof of alien existence. No one else has what you have. No one has ever had what you have. Possibly no one ever will ever have what you have. The UFO community is widely regarded as a joke by many people or just a bunch of guys whose imagination is running amuck in the swamps. A little too much moonshine and some festering crayfish meat. And you also know it's all real and would really like to help put down all the skeptics. Now hold on one minute. You are a skeptic, so you have a pretty good, down to earth picture of what's really going on. You really don't want to believe that badly. You are also now beginning to hear voices and see strange images in your head. You have no time to pursue any of it. What would you do? I know what I did.

I gave notice and quit my job. Goodbye 70k/yr! Hello my alien friends.

Oh well, I needed a vacation anyway. I was lucky to have saved just enough money to live on for a while. I figured that they wouldn't repossess my car or house for some time. I would explore the wild unknowns of aliens on earth, make some new weird friends, try to represent the human race properly and write this damn book which is taking much more time and effort than I ever thought possible. I also planted some grapes and learned to make wine. Life is strange but good alone here on the ranch with my alien friends. Invisible alien friends no less! I think we agreed at some point that if they could stay camouflaged, it might be best. No need to scare the neighbors. Anyway that's how it went.

Seriously. As far as the living conditions, I gave them permission to stay. Aside from that I'm not at all certain they are actually here at any particular moment. I have no way to keep track of them, lock them in or keep them out. I suspect they come and go as they please. I can rarely see them but hear them quite often. I have no idea where they sleep or if they sleep. They are very kind to me as I am to them. They like my crazy sense of humor and Sparky has actually gotten to the point where he can construct a simple funny joke. I suppose they like it here, or they would have left long ago. They also seem to think I am in some way special or important to them. Whatever.

Now, getting back to the communication process. I've already explained that the process was extremely difficult to understand and that I still don't understand. It started as strange sounds which had no connection to the house, activities outside, or anything else in the neighborhood. I know because for a while, I raced around the house and out into the yard to see who was making the noises. I purchased an electronic sound enhancer to bring ultra quiet sounds into my human auditory range. I even tried experimenting with simple directional objects like plastic pipe held to my ear to find the origin of the sounds. I did find a lot of birds hiding in trees, and a few critters under sheds and bees and bugs and landscaping crews blocks away and also barking stray dogs held captive in the corporate yard awaiting transfer to the next city over by the animal control officers. All from the relative comfort of my lawn, whether front or back. But I never found the origin of those noises. Yes, those strange in my brain noises. After doing this for months, I finally came to the conclusion that there were no noises to be heard. The CIA having given up on the LSD thing might be using ultra high tech devices, which could only be heard by me and no one else, to intimidate or dishearten me enough to change grocery stores which I regularly shop in, but I had no other inkling of why they would be doing this. I've heard that they can point a block shaped thing at your head and make you hear anything they want you to while no one else nearby hears a whisper. Oh, now maybe that could be it. Nope, no one in my line of sight and unless they have a miniature flying drone following me in and out of the house, I don't think that they could keep the secret audio beam trained on me either. They could be using a spy satellite to microwave beams down directly onto my property, converting those microwaves into speech as they pass through my skull but then again I don't really think I'm important enough to justify that much of a black budget. Maybe homeland security has fitted high frequency devices onto all the telephone poles nearby, and they transmit continuously and mercilessly. The Navy could be using their ultra low frequency devices to communicate to me, but they usually only bring that out to talk to submarines, not me, not just one mostly unimportant human. No, there were no noises to be heard. This left only one or two possible choices. Crazy or Telepathy.

I chose telepathy. Better for me that way.

I'm fumbling around one day in my house, looking out at the yard, and also trying to come up with a plan to re-trim some of the windows outside as well as give the house a complete coat of paint, when someone in my head says. "Hello". You must understand that I am actively thinking of window measurements when this happens. I pause and notice that there's no one around. Nowhere. Not even close though the voice seems to be right next to me. It also didn't seem quite right somehow, because it appeared to be the voice of my best friend from my school days speaking. Since that was many years ago, why would he still sound just like my old friend from school? And where was he? Why was he here three thousand miles away from the point I last saw him and why hadn't he aged? At least his voice hadn't aged a bit. But there was no one there. At the same time, an image of somewhere promptly displayed itself for a second in my mind. The image was definitely not of where I had seen him last, and I have never identified it. It passed much too quickly for that.

So now we can add voices to the list. Formerly it only included, lights, orbs, vibrations, glowing rings, strange non-human persons, weird repetitive images, generally strange stuff happening and of course, photos of alien beings.

Oh please, what's next?

Chapter 27

Into the Telepathic Dreamland

I was living in a small town near Princeton, New Jersey. Einstein had passed away several months before I was born. I am now seven years old. I am sound asleep one night. I am awakened by a strange dream. I have never been awakened by a dream like this one before. The dream is so vivid that I would swear I was living it, when I was swept back into my normal reality. In my dream, pretty certain that I was minding my own kid business, having fun running about, I suddenly become aware that a Tyrannosaurus Rex was chasing me and my brother through Princeton University. A typical kid dream. But as far as I know, I have never been to Princeton University so far in my short kid life. I seem to remember the arrangement of the buildings and the layout of the streets. The T-Rex appeared from around one of the buildings as my brother and I were running down a street. We were chasing a small red ball which bounced steadily before us. We did not seem to be getting closer to it but ran even faster when the T-Rex appeared. Giving up on the ball and loosing track of it we ran to a large building where my brother, who is a few years older than I am, did something so like him that I was immediately pissed off that I had not thought of this first. He dashed in front of the building which was somehow turning into a storefront and stood at the doorway perfectly still, transforming himself into an iconic image of one of those old cigar store Indians. As I was also near the doorway, I ran inside and found a pile of boxes against the far wall. I dove underneath and peered out from between the cardboard flaps. The next thing I remember is the Tyrannosaurus Rex walking into the building.

Step by step, inch by inch it moved closer to me. The T-Rex had apparently passed by my brother without bothering him and was pursuing me. I could feel it's hot putrid breath stream through the gaps in the boxes as it searched. The exhales and snorting from its nostrils grew closer and closer as the last box was shoved away. I opened my eyes and to my alarm was looking directly into one of it's eyes. One big red eye. But as I looked up, one of the boxes to the side of me tipped over and fell flat allowing a beam of light from somewhere outside to strike me dead in the face. I woke up. I had also peed my bed, which is probably why I remember that damn dream to this day.

Scared the living shit out of me. Well, almost.

Now I am thirteen, and I awaken in the dream of a city. I am naked and I am alone, and I walk down the street looking keenly at the buildings I pass by. I am amazed at the walls of the building next to me as they are glowing bright colors. I reach out and touch the wall and find that it's smooth and warm. I realize that I may be dreaming. It feels like the "safe place" I go to sometimes to think, but is different in many ways. The colors are so vibrant. I check to ensure that I am not creating this illusion in my mind by squatting down and touching the walkway beneath my feet. I feel the cool roughness of the surface on my fingertips contrasting to the smooth warmth of the wall. Everything is very solid, very real. I feel a breeze on my body and face, and it smells slightly of some kind of flowers in bloom. I look up and see the skyline and it is nothing like any city on earth, for each building has it's own glowing panels of varied colors. The sky itself is dark, but there is a large glowing star in the sky. Larger than our sun, it sits in perfect view and radiates it's red, amber light. I begin to walk towards the opposite building where I can see the horizon beyond this city and notice that I am now tilting at an unusual angle. As I begin to tilt further, feeling no shift in gravity whatsoever and neither sliding or falling I suddenly sense a presence near me. It's a slender male with shoulder length blond-brown hair. I can just see him out of the corner of my eye as I try to regain my balance. I am not actually falling, but my mind is confused and off balance and I drop to my hands and knees on the walk. He wears a long white robe of some sort which rustles slightly as if a breeze were blowing against it. He carries a staff. He has something in his other hand as well, but I can not make it out. He does not move and does not speak. I attempt to stand and I raise my hand palm forward motioning to him, but he does not respond. He seems to look at me as if I am strange or out of place. I see his eyes now, and they are blue. I feel as if I am turning upside down and I open my eyes wide. I feel as if I am grabbed by the ankle and pulled upward quickly into the sky with my arms and free leg flailing about. I shut my eyes, and I feel warm again. I am now awake and find myself lying in my bed. The bottoms of my feet are very cold.

For whatever reason I am now driven to paint the walls of my bedroom! I can not seem to get those bright images out of my head and I'm obsessed with recreating them somewhere. My family owns a large old house in Western New York. This house, like many old homes in New York has bedrooms upstairs off a central hallway and then even smaller bedrooms beyond which can only be accessed from the front rooms. Bathrooms are between two front bedrooms and have a door on either side. I take some money from my savings, and I ride my bike to the store where I purchase the paint. I do not know consciously what I will paint when I start. I buy several quarts and pint sized cans, all of odd colors. I have no artistic training at all, but I still try to capture the strange beauty of the horizon in that odd place where I have been. I can not even conceive of how to draw a building correctly, so I pass on that thought. I paint the longest wall of the back room behind my main bedroom black. I paint tiny stars and large planets in the sky and then add the most prominent feature, the large glowing sun. The rays of this sun are streaming out in all directions. I paint with florescent colors. On a side wall, I paint the man in the robes. My parents are horrified. I'm grounded.

Now it is the year **2009**.

I dream of electric places, mechanical places where everything is smooth and warm. There is no rumble of machinery and very little detail defining the extent of it's workings, but I know that it is a machine. This is no journey to the interior of an old Swiss watch. It occurs to me that this is like being in that place again. My place. I look around the room that I am presently in and I see vertical walls of white or almost white. A slight bluish tint is evident or maybe it's just in my mind. The machinery makes no noise, but I can feel a vibration or very faint hum coming from somewhere. It's so faint I again question if it's real or imaginary. It's coming from everywhere at once. There are points on the walls which look like notches or seams, but I am not that interested in those because there are other people in the room where I am. They don't speak and seem not to notice me. They sit quietly on a sort of chair or stool, and there's a platform or desk in front of each. This place has some resemblance to a school classroom but none which I have ever seen. In the front and rear of the room are what looks like large windows. I focus on the one at the rear. It appears to be approximately six feet tall by twelve feet wide. There is darkness there. It could be a large display but is inactive at the moment. It could be a kind of writing board. Whatever it is, I cannot see into it or through it. The people just sit, never move, and stare straight ahead. They look calm and peaceful, wide awake and somehow happy, but they do not move. I next realize that I too am now sitting at a "desk", and I wonder if I can move. I do not remember sitting down, and I am very curious at this point. I want to explore.

I have been here before. I'm certain I know this place, and I know this feeling. I think I am in the other place. What I have always called the "safe place". The place I used to go as a child and learned to explore as a teen. This time though is again different. I have never encountered a person here before except for the being in white robes. This looks strange to me. I am going to explore. I stand.

Suddenly there is a feeling of tension which flows across my head, chest and upper arms. My attention is drawn to a point above me, near the ceiling. There is something bright hanging there. I don't believe that it was there a moment ago, but I was not looking up then. I suspect it's the ceiling light, but as I watch it, it turns red. Or was it red to begin with. I'm not sure. The funny thing is that it doesn't shine onto any surface. It just hangs there glowing. Calmly I step away from the "desk" and once again look around the room. No one has moved. No one has looked in my direction. No one else has stood, and no one has spoken. They all still sit pleasantly staring into space. I begin to see and take note of interesting items such as that window or whatever it is on the wall.

My mind suddenly feels relaxed but foggy. I decide to look around, and I see walls of bluish white. If there's machinery, it makes no noise, but I can feel a vibration or very faint hum coming from somewhere. It's so faint I again question if it's real or imaginary. There are points on the walls which look like notches or seams, and there are other people in the room where I am. They don't speak and seem not to notice me. There are five or six, and they sit quietly on a sort of chair or stool, and there's a table or platform in front of each. This place has some resemblance to a classroom but none which I have ever seen. In the rear of the room is something like a large window. It looks to be six foot by twelve feet wide. There is darkness there. It could be a large display, which is inactive at the moment. Whatever it is, I can not see into it or through it. The people just sit. They never move and stare straight ahead. They look calm and peaceful, wide awake and happy, but they do not move. I next realize that I too am now sitting at a "desk", and I wonder if I can move. I do not remember sitting down, and I am very curious at this point. I want to explore.

The feeling of deja vu floods over me. I have been here before. I know this place, and I know this feeling. I am in the other place. What used to be my "safe place". This time though is different. Something else is here with me. Something is watching me. I am suddenly alarmed. A shiver rolls down my back. I have never encountered anyone here before who interacted with me in any way if you exclude the being in the robes. But I have an alarming sense of presence now, and this all looks strange and somehow different to me. I am going to explore. I am going to stand. I am going to go.

Immediately my attention is drawn to a point above me, near the ceiling. There is something small yet bright there. The feeling of deja vu is much stronger now. I don't believe it was there a moment ago or was it? I suspect it's some sort of light, but it's red. Was it there before? Has it always been red? Yes, it's always been red. I am sure of this now. The funny thing is that it doesn't shine onto any surface. It just hangs there glowing. Calmly I step away from the "desk" and once again look around the room. No one has moved. The red light is still up there and pulses briefly. No one else has moved They all still sit pleasantly staring into space. I begin to look around the room and take note of interesting items, and as I do I remember my self-training from so many years ago. I focus. I will myself to move. I will not do this again. No more tricks. I will explore.

I begin to walk around the room. Nothing happens out of the ordinary. The feeling of alarm subsides to some extent. Safe so far, but who or what is controlling this place. It's not me. But now for the time being, I am in control again. I am also now aware of the rules of this place or at least the rules of my safe place which for all practical purposes seem to be the same. I approach one of the happy silent ones and touch her shoulder. She is around twenty years old, though I am not a good judge of age. Her hair is short, and her face looks vaguely familiar and masculine in some way. I would have suspected that she was a he but for the protruding breasts and obvious bra lines under her shirt. She does not move. I gently wiggle her shoulder. Nothing, no reaction. I notice that her clothes are stylish, and they are clean and neat. I leave her side and move to the next table to the left. There sits a boy about fourteen or fifteen years old. His shoulder length reddish brown hair falls across his forehead as he stares blankly forward. He is thin and wears only shorts but no shirt. His feet are bare. I run my fingers down this arm. He does not move. I leave the silent people and move to the rear of the room where the window or screen is somehow attached to the wall. As I approach to within about four feet, it immediately lights up and displays what looks to me like three dimensional art work. Strange areas of colored light with some sort of lines connecting them. Seeing no apparent meaning at all to this, I turn to see that the wall of the room is open in one place. The room itself is circular or possibly oblong in shape and seems to wrap around itself somehow. In the overlap on one side near the front, there is a doorway. I immediately move in that direction. The overlap of the room forms a hallway of about twenty feet in length. As I walk into the hall, I see an open door at the other end. As I get closer to this door I begin to see that there are niches about two feet in front and on either side, near the doorway itself. They are about three feet wide and six feet high. As I approach, I begin to wonder what could be on the other side of the door. I notice that as I move closer to the opening, the light in the hallway is becoming a deep shade of red. I can not see very far into the next area, but it is lit with bright white light.

Suddenly I see something move. It's there in the shadows near the doorway. I can just catch a bit of it from where I stand. It moves like water rushing over rocks in a stream or the leaves of a bush rustling in a breeze. It's slips into one of the niches there and is mostly hidden from view, but I catch a glimpse of it again, and it steps forward. I freeze. It looks vaguely like a large dog, but no dog any human has ever seen before. Nothing earthly ever looked like this. It's size is like that of a large Saint Bernard but what looks like hair is more like the wavy fur on my dog Greggor, only alien and surrealistic. It's as if it is not hair at all. It moves on it's body like leaves rustling. The hair looks wet or oily, but it does not drip, and there is no smell at all. The individual hairs almost seem alive and sometimes move together in groups and sometimes individually but always continue to move. The hair seems to sometimes point at me. It's vaguely animal eyes shine with a piercing red light. It seems neither mechanical nor robotic and moves as if sliding through space. It moves forward and blocks my path. Then I immediately feel my brain hurting like I am having one of my migraine attacks. A sharp dagger of pain shot directly into the sides of my forehead. It's the way the light wavers and glints off this thing's body, which causes the pain and gives me that pressure feeling in my forehead. Where is the light coming from? There's no other source of bright light in this place! The light seems to emanate from this things body! It's as if someone is shining a spotlight on it and reflecting it directly at me. I once again muster my courage and focus. I clearly see now that this thing, which looks like a dog, is not a dog at all. Not from my reality anyway. It exudes a monstrous deadly quality which is beyond description. It's horrifying! It looks up slightly and meets my eyes. The glowing there burns and flickers but does not shine outward and is all contained in those tiny orbs. The mouth opens, and it snarls. It's fangs, and teeth glisten though there is no light directly on them. The slickness of saliva seems to drip from them as the lip continues to curl upward. I begin to fade off, but again I refocus my mind. Again and again it steps forward, but I do not retreat. I will not retreat. I know this place. I understand the rules here now, and no vicious looking simulacrum of my dog Greggor is going to intimidate me into going back there, into there. Into that room.

It stares a deadly stare at me. One which says, "I am the

beginning and the end" "I am your death, and I await you."

I take a step forward.

His mouth opening wider, teeth bared, he takes another step towards me.

I put one knee on the floor and hold up my hand to him as if to say stop!

He moves to within striking distance.

I withdraw my hand, bending my arm at the elbow, but only part way. I do not want to give the impression that I am afraid, and I am not. I stay focused. My hand still outstretched, the beast stops and tilts its head to one side. It turns and cocks it's head slightly to direct it's hearing towards my location like a human dog using every trick to determine my movements and predict my next action. It's ears are laying back yet pocketed forward to catch every tiny sound. To hear my breathing and be ready to pounce when I move. But, there is no sound. I have not heard a single sound since arriving here. I suddenly realize that there *is* no sound at all in this room, in this place. And so I immediately realize that this must be some sort of trick of this thing, this death dog. I suspect it's nothing more than a doll, a stuffed puppy in a scary suit designed to intimidate all those who might awaken and attempt to leave before they are released. It is clearly imitating my dog. It may have some defenses and may use them, and it must have known about me in advance or instantly found in my mind the worst vicious confounding and disconcerting beast of my deepest thoughts that I could ever imagine. My own dog turned rabid, alien and deadly. Nothing could shake my focus now. This was it, all or nothing. I wasn't afraid. I was calm now. I knew exactly what to do.

As he moved forward again snarling, I drew my hand back up towards my body. I waited. The beast moved close enough that it could have lunged and ripped my throat open in one split second. Still blocking my path, it hunched down with both eyes locked upon mine, inches away, but it was now within my reach. I did not move suddenly or quickly, but I raised my hand with the palm facing this thing and in full view of those glowing red eyes, I slowly moved my hand past the razor sharp teeth and gently rested it on the side of it's neck. It's teeth still showing, it sat completely down on all fours on the floor, and I slowly stroked it's head and body. The hair feels like liquid metal goo, slippery, but cool to the touch. There's a small continuous electric shock, as if some static electrical charge was escaping from it. It closes it's mouth but still displaying a snarling expression, it looks up at me quizzically.

I slowly stand up and move forward and only briefly look back to once again see it's eyes locked upon me. I keep walking. It sits and watches as I move through the doorway and exit this place. The last thing I remember there is that, for some reason, I paused for a split second, only long enough to bow my head to him slightly as I stepped backwards through the doorway.

Chapter   28

Into the light

I am immediately in a very bright place. In the smallest fraction of a moment, I set a point and imperative in my mind so that I will not forget everything upon returning. I wonder if this is a dream or something else. It has components of dreaming and of being in that other place and also something very different, a new puzzling piece. I have found or have been found by a new entity which inhabits this strange world. I wonder if I have found a new friend or at least earned the respect of this new being I've encountered.

In the middle of this thought, a voice sounds off in my mind and states that I have made it through nine doors and that so far I am the only human to have done so. I answer, "What doors?" "What were they?" "Is this a test?" but there is no reply. I ask over and over again, but there never is an answer to these questions. The only reply, "There are twelve." I think I remember some of the tests. The first, in the year 2000. One night as I am lying in bed reading, attempting to relax enough to get to sleep, I set my book aside, turned off the light and prepared to dose off. But lying there I begin to become aware of a presence in my room. What I mean by this, I am not exactly sure. It's that feeling you have when someone is watching you although you can see all areas of the room and no one is there. Then a thought in my head asks me to close my eyes and attempt to make a bright blue sphere appear in the darkness inside my eyelids, inside my mind. Voices in my head? This is before I had any recognizable contact with the aliens. So, OK. I'll bite. But colors? Why colors? Has anyone ever seen colors in their closed eyes in a dark room? Maybe. I don't know. Why should I attempt to do this? Then I started thinking. My subconscious certainly didn't dream this one up. So I closed my eyes and stared into the darkness. Little white lights and blobs appeared but no blue sphere. I heard the voice say, "Try again." So I relaxed and concentrated and in the blackness and to my amazement appeared a blue sphere. "Again"! The voice said. I did so immediately and produced another blue sphere. Then the voice asked for a red sphere. Then a yellow sphere and finally a green. I did this quite quickly, and the voice said, "Thank you" or I felt it think, "Thank you". I am not entirely sure of this either, but I immediately fell into a very deep sleep. I must have. I awoke in what seemed like moments later, fully rested as the sun peeked through my bedroom window. It was now morning. I only remember approximately two or three minutes of time passing.

I sometimes ask myself whether this is all actually real. I sometimes look in the mirror to see if I am who I think I am. I do this to reground myself to basic human reality. So much of my thought process these days has become attempting to understand just what is happening and what these aliens are doing here that I sometimes loose track of that essential substance which makes me a sane human. I loose sleep. I eat too much. I don't exercise enough. I'm grouchy, and I......well then again maybe it's just old age catching up with me. Or maybe my brain is just struggling with that technology so advanced that it does seem like that magic thing again. And I sincerely wish to see through that magic because it's very troubling for me to be in someone else's magic show that I think should be my own. Am I in some weird sci-fi movie? Am I now a stranger in some strange land? A lost Magi alone in the wilderness? Will I find some ancient tablet and start chanting the words inscribed upon it, to suddenly find, Merlin standing before me, looking magnanimous, and shouting something about the dragon's comings and goings?

Deeper still..... way deeper.

Chapter 29

Feathered Dragons and Winged Serpents

Light.

Thought

Almost surrealistic images.

Dragons. Ancient Chinese writings. Dragons from medieval times. Feathered Dragons!

Visitors from the time of the Mayans! The feathered serpent god Kukulkan.

The Mesoamericans called him Quetzalcoatl! They say he often traveled with his companion

Xolotl, a dog headed god.

The one who escorts and protects. The one who comes with the sun. The one who stands guard, like a human's guard dog. The bringer of peace. The being who travels to unknown alien lands making contact peaceful. The dragon who is wise.

These alien minds just love multiple dimensions, actions and coincidences. They're absolutely manic about it. Go figure. Whatever. Eleven dimensional lucid chess dreaming.

I look for similarities between our photographs and experiences and real life. I say "our" photographs because there are definitely at least two intelligent race's influences there. If the bipedal being "Sparky" has been returning here for many years, he may have influenced many culture's dream imagery. His companion, the feathered dragon, travels with him and seems to act as a scout and body guard. In ancient Mayan culture, he appeared as Quetzalcoatl, the feathered serpent. In Chinese writings, he appears as the winged feathered dragon. The serpent has a prominent role in many ancient religions. Adam and Eve were supposedly tempted by a serpent in the garden of Eden. Serpents were worshiped by many ancient cultures including the American Indians. Stories of serpent gods date back to the earliest religions. He also appears in our photograph. He coils about Sparky's legs, and his head rests between Sparky and Greggor. Another of these beings is sitting next to Greggor in the grass. It has a dog's head and takes on an eerie likeness to Greggor shifting it's head to the same degree as the photo is taken. Could the vicious looking Greggor simulacrum from my vivid lucid dream-like experience be this being? I believe so.

Was I merely lucid dreaming? I believe not.

The feathered serpent. The one who protects. Sparky always keeps him close. He always keeps him between himself and the aliens (humans). He surrounds and protects Sparky. Sparky seems to climb out of his coils. He travels with Sparky. It is written that he sometimes travels with a feathered serpent dog. He flies. He sometimes teaches people new and wondrous things.

He most times looks friendly. He can also look intimidating but has never harmed humans. He never asks for human sacrifice. Never. He does not wish to be worshiped. In ancient times, the Mayans revered Quetzalcoatl as the god of earth, water and vegetation.

Then, suddenly some strange images float through my mind.

I recall all the activity and apprehension surrounding the ending of the Mayan calendar and the possible end of time in the year 2012. The end of the current and last Bak'tun that the Mayan priests wrote down for their people. The end of time itself. The end of the world. There were so many documentaries and books that every one of us knew of the coming event. Every one of us knew that we could do nothing to stop it if it were actually to happen. The Mayan priests started the Mayan calendar on 3500 bce, thousands of years before their culture rose to greatness. They ended it on December 21st, 2012. There are stories amongst not only the Mayans but also among other cultures that this is the time when their gods would return to walk the lands and judge the living. If the land were not cared for and if the living were judged to be unacceptable, the world would be destroyed. The Apocalypse.

Could these beings be the same ones predicted to return? Could they have returned to see our progress, to view and understand our new cultures and determine whether or not to allow us to continue to live on this world? Could this actually be them? Could the being with his symbols of high rank and feathered serpent allies be here for something besides just contacting the human race? To judge us? Could he possibly be the harbinger of our fate? Could the dragon be dangerous?

Ultimately, I have found him to be peaceful and also to be our friend.

Feathered Dragon.

Chapter 30

Hints Concerning Reproduction

I still have that damn photo. Yes, They are still here. Those beings we were discussing, I think I also have a sort of feel for them now, as well. I can tell you now that from what I have learned, the race is physical in nature, and in their own way, intelligent, creative, inquisitive, funny at times, and also creates youngsters just as we do to sustain and grow their population.

If we look back again at the Pioneer and Voyager spacecrafts, we see that we humans did not really put a lot of information in our messages there. If an alien race found and studied the plaques, they would see two individuals with two different types of bodies. We immediately see these as a man and a woman, our two basic reproductive types, but how would an alien race interpret them? There is nothing to say that these two humans interact in the reproductive process. There is nothing representing the result of a reproductive act, such as a baby to show them why the two individuals are there together. How would that work anyway? If an alien race, which may not even have a separate male and female, or something which looks anything like us, should see this, they may think that the intelligent life we have on the planet consists of two different species of humans. One has a growth hanging between it's legs. Not a very good place to have something like this if you're to do much moving about or running, so it's most likely the one that gives birth. Maybe it buds off it's young similar to some plants, and this is the start of an offspring. The other has large chest muscles and hips, most likely the worker type. Do the two have anything to do with one another, besides living on the same planet? I'm certain that there were only good intentions put into the design of the plaque, and it was a difficult task to communicate facts to someone or something when they had no idea of what the heck they would be talking to. I know I could not have done it any better. So what was NASA communicating? Well, I guess the answer lies in the map to earth. The message is "Here is your map to our planet, and when you get here, this is what you'll find. We are the ones who created this machine, and there are two types of us." We see them as a man and a woman.

And, Mars needs women! Just kidding. I have no proof of where these guys came from, and there is no indication that it was Mars. However since there are no women on Mars that we know of, I guess it is correct to say that Mars does need women. The photo shows a larger alien and a smaller one on it's back. So in tune with my past comments, could the smaller alien be something other than a child? It could! It could be a small mate! Along those lines, there generally is some difference in size between males and females on this planet and some are extreme. The female golden orb web spider is ten times the size of it's male counterpart, which could be a bit dangerous especially if big baby is a bit testy the day little mister horny comes by to do his stuff. This theory is shot down by the fact that the little alien seems to be wearing something around his butt and between his legs. OK. So a mate might be incontinent, but let's get real, would you bring an incontinent mate to have your picture taken representing the home race? I say no, not even an alien intelligence would do that. Maybe it's a condom!

The little guy could be in the middle of doing his thing and just dripping with male fluid, but I would think they would want a little privacy for that. The smaller alien could also be a "bud" or clone in progress. They may not have two sexes at all and reproduction consists of cloning oneself through the practice of squeezing off your little bundle of tumor joy and letting it develop on its own. It's entirely possible that the smaller being is a mate just hanging out on the females back, but if that's so I would imagine that it's simply for the photo opportunity because it would be a strange thing to do otherwise. Oh, but then there's that diaper thing again. Maybe it's just shorts. Or something to keep his little thing in. Maybe he just needs a ride since he's smaller. Possibly, he will not grow larger, and has evolved a symbiotic relationship with the other who carries him like a taxi cab to wherever he needs to go. We have had no verifiable direct contact with alien races until now, so there's no way to tell exactly how they reproduce, but my guess is that, however they do it, the little guy is not a mate but is in fact a bouncing baby alien.

Possibly the larger alien is a female? Again, I'm not sure, but it certainly seems to look as much like what I would think an alien female might look like. Possibly pregnant? I think, that having been said, we encounter the same questions we stumbled over before. What if there is no male counterpart whatsoever? What would that look like? We certainly have not seen anything like human genitalia. Where would an alien hide such a thing? Inside the body like our earthly whales and dolphins maybe? Perhaps it only protrudes when necessary? Possibly, or else it's mate is something which we haven't seen yet. Maybe a large green bug eyed monster of a mate!

It's entirely possible that the photo shows both a male and female alien and a child. How can this be? Because I believe it possible that an alien could change at times between the male and female sexes. Basically, that would make it both androgynous and hermaphroditic. It's always both male and female by nature. It's male when needed, and when the urge to reproduce asserts itself it becomes female and produces a child. So could it be a male anyway? Where would you hide your male genitalia? Maybe nowhere! Possibly it has none. Or does it?

The rope or staff or wand is one of the most interesting items in the photograph. Is it possible that this thing is actually a biological extension of the alien's body? It does appear to be attached in some manner and looks vaguely biological in nature. What if this thing is a male's reproductive appendage? A very long penis? It's possible that since we showed the body parts which humans use in the act of reproduction that they would do the same. Let's assume that it is attached to the body and that this long thing is originating from the adult. Could it be that this extension of their bodies is a way to mate? This of course would mean that our friend is actually a male! I'm not sure this works with the information presented previously unless the males rear the children! So I will stick with my former theory for the time being that this entity could assume both the male and female roles.

It's time to look elsewhere to find out what our rope or staff may be and whether it's a he or she.

Chapter 31

Looking Elsewhere For Clues

It perplexes and annoys me that I can find many earthly references to staffs, scepters and magic wands but few descriptions of why they were designed as they were or used in the first place. We should probably go there next and explore these myths, stories and old references. But there's also another side. This other side of the thought process will later take us into the darker side of life itself. The place where monsters live whether alien or earthly in nature.

A petroglyph is a rock carving, and a pictograph is a painting on rock. Many can be found in the American southwest where ancient American people scratched or painted the important events and characters in their lives onto stone for preservation. Some of the images are thousands of years old and depict what looks like other worldly beings. The Indian people speak of beings who helped them survive cold, hard winters and dangerous events in ancient times. The rock carvings and paintings of Utah, Arizona, New Mexico and California depict beings which seem human in some instances but in others look strangely similar to the ones in my photograph. If you look at the sketch of the image from Sego Canyon, you will see an odd being with a strange set of ears or streaming energy, being emitted from its head. There are many such drawings from the same location and some show halos or coronas. In this drawing by early humans you will also see staffs or sticks which appear to have no purpose. They are not spears because they are all wild and curvy in nature. What could they be? Walking sticks? Herding tools? Why would there be similar items to the one seen in my photograph of aliens? I don't know, and there is no explanation recorded in writing, but since ancient times many of these items have been painted onto rocks or carved into them. The early Indian people depicted important events in this manner. Important events to be remembered. Important animals. Important people. Important spirits, as they called them.

The same staffs or curvy wands also become entwined with and are documented in later social structures, stories and works of art. Lords or rulers have carried scepters or ceremonial swords and have worn crowns for thousands of years. Coincidence again? Possibly. But is it possible that we have over the years developed a reverence for beings which appear only fleetingly in our minds and just happen to carry or wield a staff? Is it just our fear of someone who carries a weapon or something else? Something deeper. The old man who suddenly appears on the road near the woods. He carries a long staff made of curved wood. He speaks in riddles and then disappears back into the shadows. Maybe it's a curving staff, and perhaps it's not carried but attached somehow, but it seems that we have a healthy respect for those that have one, as well as a crown.

Sego Canyon Pictograph Impression

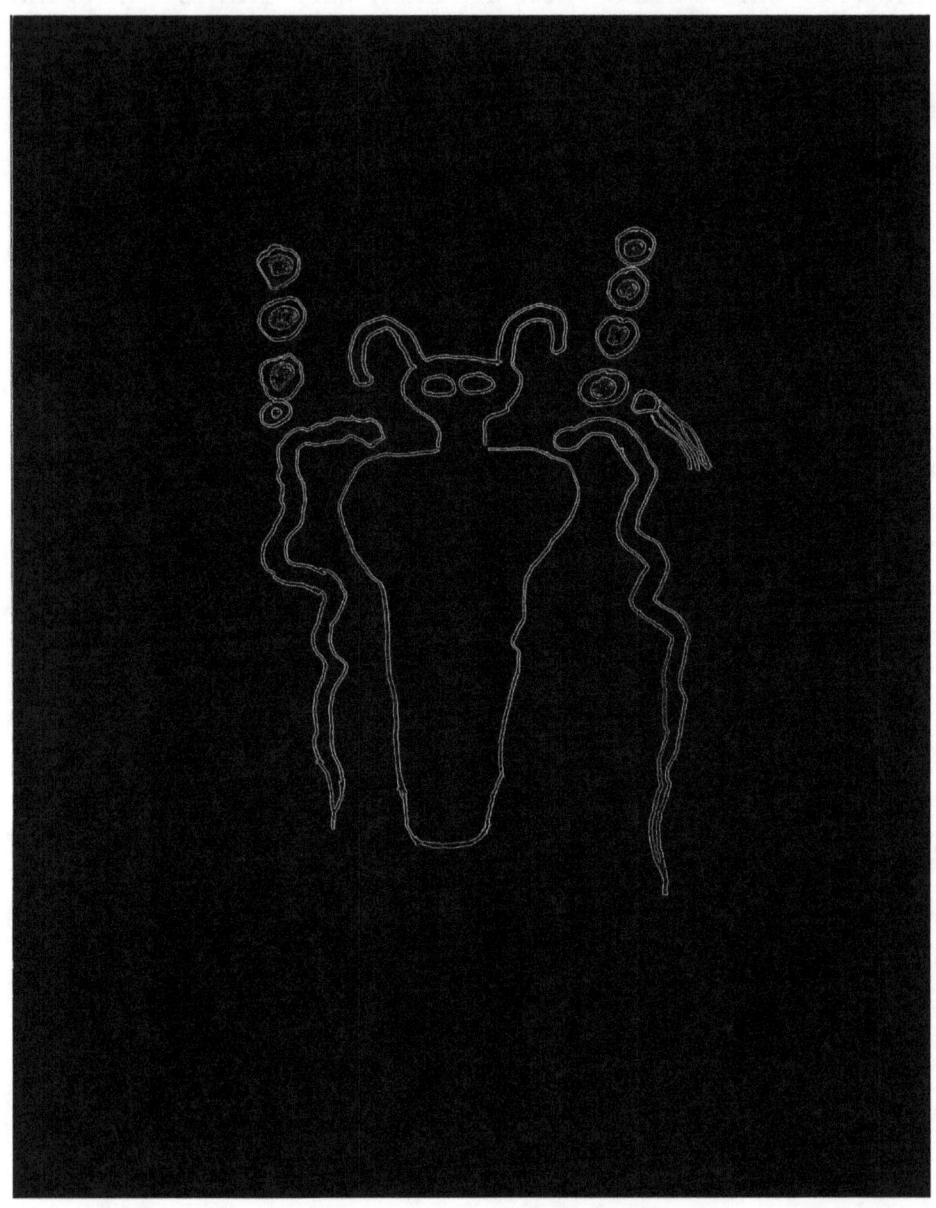

The fact that Kings have always carried such an instrument of respect should tell us something. Power. Majesty. The right to lead the people, having been given to them by the gods or God if you will. The holding of both attests to their status above mere mortals. They stand in that place where they can lead their people but also speak with their gods. The place between two worlds. Somehow we know this to be true, don't we. We also place them higher than all their subjects whether seated or standing. Subjects bow down in their presence and lower their eyes. Don't mess with the one who carries the staff and wears the crown. These objects are enough to say something of importance in themselves. If you cross me, I shall use them against you. My power is far reaching. I speak with the gods. In one ancient religion, a shining entity with a sword and adorned with a golden crown is always depicted with a bull as if the two were united in some fashion. According to Mithraism, the sun god Mithra is forever shown holding a sword and sacrificing that bull. This could be quite important. Did our ancient ancestors make a conscious decision to transform a rather strange looking alien from who knows where into a nice appealing human form for the general masses to worship? You must admit looking at my alien, he does have many features which resemble a cow or bull. Especially the face and head. To ancient people the spikes on the head might look like horns, the large protrusions on the side could be interpreted as the ears of a bull, and the face looks rather bovine in total.

Might the horned bull image have been symbolically sacrificed to make way for, the more acceptable, image of Mithra, who now forever appears in human form? The bulls face becomes human. The bull's horns become a gleaming crown. The sword shall forever be an enigma.

The magician's wand might be thought of in the same light. Why again would a staff of wood hold power for anyone? First though, what is the definition of a magician? It's the man who uses the power of magic. He uses the magic wand. A Magi, a Merlin, a wizard, an enchanter. All names for the same thing. In ancient times, a wizard or sorcerer was a man who had been schooled in and was proficient in the mysterious and important arts of the ancients. He knew how to manipulate reality for his own means and desires or the desires of his king, lord or leader of a tribe. This knowledge had been handed down for many hundreds or possibly thousands of years to him, and no other except the other men of his clan or group knew of these ancient arts. Think about it. A magician is one who suspends our reality only to inject his own for a short period of time. And as our reality is suspended he can then use this time to communicate the changes he feels will perplex, tantalize and amaze us, only to return us, moments later. You may have heard another definition, but I favor mine and it also in some ways describes these alien beings we are discussing. So back to the magic wand. Why did magicians always seem to need a piece of polished wood to represent their ability to suspend reality? Are the alien beings the inspiration for this? The classic or ancient magician always carried a wand or staff. While the human staff is the most ancient of all like items, it is first described in writings as a shepherd's staff. Could this staff, which our alien friend exhibits, be some kind of herding tool? Greggor did seem to want to seat himself in an exact spot for the photo. Could he have been coaxed into that position? The being seems to be setting it's wand out into a curve which would lead directly to him. We may never know, but the relationship between the alien rope-staff-wand thing and Greggor makes my mind reel! Could it be something else besides a herding tool? A spacial link to our reality or dimensions?

Perhaps a wand to set a point which defines this being's space or even his time and separates us from his reality so his devices can mold his area as defined by the circle, occupied by the feathered serpent, into a piazza which extends to that point, so he can stand and communicate with us for a time?

Possibly a mental connection to our reality, as well? In other words, could he be speaking through space/time utilizing a connection we have never even imagined could exist? A physical, "biological" link which allows him to inject his thoughts and desires into our reality. Possibly even into the minds of humans and animals alike? A carrier wand to transmit thoughts out of his dimensional space and into ours! OK. I think I'm getting a bit off track here. The device looks biological and so I hesitate to grasp this last idea too firmly. So let's keep moving on, though we are now running out of side streets on which to tread. Many have been dead ends or have led to doors which I have hesitated to open. We have so far left them locked. But we must open one now, and it leads to that dark side.

I have searched my mind and soul to find some way out of this, but I can not. I must make you aware of the possibility of dangerous behavior by these things and yet having known them, I will later, I hope, bring the issue back into perspective. I am torn. I do not want to scare anyone or to create an atmosphere or fear and yet I must go forward with this thought. It rakes over my soul like a large lizard's claws and rips deeply into my flesh. I anticipate the pain and ooze of blood weeping from the self inflicted wounds, and I run from the thought. And so with much anxiety about this, here I go.

Deeper still.

Chapter 32

Cattle Mutilations

Cattle mutilation refers to the the unusual killing of cattle, sheep or horses by unknown means and involving the draining of blood, removal by surgical means of the eyes, cheeks, udders or rectums as well as sexual organs. The incisions or cuts are so clean as to have been made by a surgeon's scalpel. The phenomenon has been documented as early as the 1960's. Sometimes the dead animals show not only complete loss of blood but also incisions which lead to the heart, which when cut open reveal that the heart has been liquefied and removed through a small hole drilled into the animal's side. Destructive to say the least! Any human would be sickened by this behavior. But what are we talking about? Predators? Hunters? Maybe even witch cults? Hunters would not be interested in shooting cattle, and predators use their teeth to rip and tear out sections of meat, and rarely leave the best parts untouched. They also leave tracks, but even in new fallen snow, there are none around the animal. The FBI and law enforcement claim that devil worshipers do not kill cattle in this manner. Their way of sacrifice is much more ceremonial and follows specific patterns and rules, leaving lots of weird evidence. In my mind, this involves lots of items which are there for no other reason but for the humans sake. Props so to speak, designed to mystify the worshipers and baffle their thoughts and minds. Feathers, mystic symbols, bowls for the blood displayed almost theatrically enhance the ceremony and lead followers to believe they are doing something of secret magical nature unknown to the bulk of society, when in reality they have simply killed an animal and bonded over its corpse.

But hey, how about insects? Couldn't they be to blame? Some insects do eat the diseased parts of animals don't they? Maggots are currently being used by physicians and surgeons to remove dead flesh from humans, leaving a clean area which can then be healed. But maggots like to feed on diseased and dead flesh. Many of the mutilated animals have died within hours of being seen, perfectly healthy. What really could be the cause?

I specifically focus on the incision which leads to the heart. A bullet might be to blame in many cases. The hole starts in the hide and extends directly to the organ. One big problem though is the lack of a bullet. Another is that there appears to be no exit wound. I then suppose a long smooth specialized cutting and feeding tool would do quite perfectly to create the holes and the removal of the blood and interior organs. I suppose that it would make our being some sort of predator. Something which needs to feed upon flesh and blood and which also has some preference for certain parts of the animals, but also does not understand that the animals are also an important part of the human's food chain, and as well as a very valuable asset to the ranchers who raise and market them. How much could one family of little aliens eat? I would assume that what would be left over after such a feeding would be just what is left when someone discovers a cattle mutilation. The similarities boggle the mind.

But then, everything must eat. This evening I felt a stirring in my stomach and left my writing to go to the refrigerator, and opening it I reached in to find a lovely t-bone steak which I had purchased at the local grocery store. Since the day was warm and pleasant, I decided to go outside and fire up the BBQ to cook the nice marbled piece of beef. I waited in anticipation for the chunk of raw meat to slowly sizzle to a medium rare. Being a fairly good cook I waited only until the blood and juices were still oozing from the surface, but the meat was nicely browned. I took it from the grill to my plate, added some of my brother's baked beans, added some salt and sat down to eat. I took no note of the fact that this piece of meat was once a living being as I cut into it with rampant desire. I ate. That's all I know is that, I ate, and it was wonderful. I do not know what happened to the rest of the animal. I could care less. I felt no guilt or remorse.

Sometimes we really do not know what the consequences are of what we do, or of what we eat for that matter. Sometimes we are not even sure what to eat and what not to eat. Different cultures have differing views as to which animals can be eaten and which animals can not. Most Hindus do not eat beef and feel that the cow is sacred. Americans love beef. I was once in another country. I proceeded into an eating establishment and met a young man there who invited me to his table. I accepted and met his friends who welcomed me to their town. I looked at the menu and during friendly conversation decided to order the same as they were going to eat, one of the restaurant specials. I didn't understand the language well and so I stuck with the crowd, assuming it would be good. The meal was presented to our table, and we began to eat. I did not realize that what I was eating was stewed dog meat. When I found out I was sickened and soon left, yet every one else at the table was pleased with the meal. I sometimes think back upon this experience when I deal with this issue.

So once again I say, "everything must eat". Stray dogs must eat. Cats, alligators, snakes, and roaches must eat. The world and everything in it is predatory in nature. Every living organism preys upon some other source of food, whether smaller, more vulnerable, living, dead, organic or chemical in nature. Even our own sun will eventually consume the earth as it continues to grow over billions of years. Black holes will continue to suck in debris and planets and even whole solar systems as they feed. This is our universe. So enjoy! Every man for himself! Go eat!

Do these aliens consume beef as we do? And being much smaller, and alien, would the blood and juices of cattle be more palatable than a porterhouse steak? How bout the tongue meat? Tasty I might say from my human point of view. A quick snack on the run perhaps? An essential meal for a starving group of beings without provisions stranded on an alien world? Speculation of course. But the pieces are slowly coming together. Are we looking at a race who having visited here for many years has landed, explored, and lost ships and crews to the hardships of exploration? Mechanical malfunctions of magical technologies? Sailors, stranded seeking food or fluids on a strange desert island of blue, spinning in orbit around a distant star called Sol?

I'm not certain what the rope-staff thing is for, but I know that this alien has no intention of doing our race harm with it. I have tried to relate to them that cattle are not animals for them to feed upon unless no other source of food is available and they will die without it. I have also tried to enlighten them to the financial issues they may burden the human ranchers with, when destroying animals. Are they actually feeding upon the cattle? Do they understand? Your guess is as good as mine, but I attempted to send the message. That's all I can do. I am not an expert. I wish I had better answers, but I don't. This having been said, the fact that we dine on beef, and they can not would surely throw me for a loop if I were an alien. Maybe the waste aspect will get into their skulls. Will he spread the message? Not known. How many of them are there? No idea. Are they all actually stranded? Probably not. But who knows. My gut feeling is that they are not to blame for the mutilations. Not directly at least.

Chapter 33

Gooey Light for Geoducks (pronounced gooey-ducks)

Having been raised in New York State, I rather often had occasion while growing up to sample seafood including local shellfish. We rarely dined on oysters, but clams were served everywhere, and I developed and strong liking for the little mollusks which were usually served on the half shell. If you have never had the experience, fresh clams on the half shell taste nothing like canned clams or cooked clams, which I do not care for nearly as much. The raw clams are the thing to go for! They have a wonderful almost sweet flavor. Add a dash of hot sauce, some cider vinegar, a little horseradish, don't chew much, just crush and you have the most excellent eating experience you could ever ask for. Add a cold beer and I'm in heaven! Oysters, you can keep em, at least wherever raw clams on the half shell are served.

When I moved to California, I was introduced to oysters but never lost my taste for clams, so, needless to say, I have always looked for specialty restaurants and places where I could get them. Today, they are much easier to find but back then it was quite challenging. So anyway, while shopping in a Seattle seafood market I came across something which just blew my mind! I looked at the seafood on ice in the case, and there was the largest clam I had ever seen in my entire life! A huge animal weighing about two pounds and having a protruding appendage which was as wide as the entire clam shell and nearly four times as long! It's meaty appendage was enormous and much larger than the clam shell itself. I found out that this was it's feeding siphon which was now retracted, but when the clam lies in its home environment beneath the sand the feeding siphon can be extended to over three feet in length! Further inquiries told me that as the siphon stretches out through the sand to find plankton and other edible goodies, it becomes much more slender. This is the geoduck clam! It is found on the north west coast of the United States. It's extended siphon tube stuck in my mind and made me realize that the rope-staff thing which the being exhibits might just be a feeding tube, and probably also a sensing device as well. Because anything which can be extended that far from one's body better be able to find it's way, or it just might get chopped off by some other hungry or dangerous creature. I'm not saying that the rope staff has a mind of its own, but merely that just as we sense touch, heat, pain, etc through our fingertips, so might this being sense similar things through it's feeding extension.

Here on earth, elephants use their trunks in a similar fashion. The elephant's trunk can not only be used to breathe, but can also be used to suck in water like a straw and squirt it out like a fire hose. It can be used to make noises, grab heavy objects like trees and logs, pick up delicate small items such as fruit or stones, and can also carry one's trainer around the pen to the enjoyment of little children in audiences at zoos and wild animal parks everywhere.

One far fetched idea is that this feeding tube could also be used for elimination, either liquid or solid or both. It does seem to originate near where our human digestive organs are located, which means that it could be both. In and out. Can't say for sure, but I bet the cryptobiologists in the audience would certainly love to look into it! I'm not sure how the cleaning of such a device would work and so I'm not a big fan of this one, I mean the idea of the tube being used for both consuming food and eliminating waste? Let's say you have something edible, like a burger. A burger makes sense seeing as how cattle mutilators generally use beef as their mutilation of choice. Choice beef that is. No, really. So there's this burger in front of you. You slip out your whoopee stick extension thing, and you feel around to find a good angle of attack. The extension has some sort of tiny cutting edges and suction device, and as we have said before, you can feel your way around with it and direct it's actions. Since it's thin at the end, you would need to liquify your meal before you slurped it up into the old alien tummy. Some sort of liquifying pre-digestive solution would be needed, and lots of blood or juices would also help. I enjoy lots of juices on my beef. Don't you?

So, you put your razor sharp chewing teeth to use cutting small bits of burger, and as they are collected, you squeeze a small amount of fluid into the chewed mass to make it easy to travel up the tube. Since the tube right now has liquifying fluid in it, it makes for a smooth process so far. When you get enough juice to suck up, the extension is already mostly empty of digestive fluid, and you work the meal back up into your body. After your meal, given a few minutes to let your burger settle you feel the urge to purge. The tube then allows waste fluids to move down and out of your body. If the waste wasn't too icky it just might work, but as I say I'm not a big fan of this idea. Maybe it's because I'm human, and the thought of using the outhouse as a second kitchen for cooking the family meal, is somehow just not right.

This might also appear to be a false idea considering that the smaller alien seems to be wearing a diaper or underpants! Clearly the little guy eliminates waste similar to the way we do it.

The larger one then most likely does too. So perhaps the rope-staff-extension is more like an elephant's trunk as we have discussed, and not an elimination device at all. This would make more sense, and the advantages to being able to explore things beyond ones immediate arm length while also utilizing other senses attached to that extension could be superior to just using something like our own human eyes and hands.

Our being also seems to be manipulating some sort of ringed or circular area of light energy. I just wonder if maybe it is using the extended appendage to sound out the terrain outside the bright envelope. Is it possible that it can not "see" as we do and uses it's extension to find it's way and know where it is on our brightly lit planet. Could the circle be a transport device and to be able to appear to me, the alien had to stall the process at an unusual point leaving it temporarily blind, and requiring it to use this extension to negotiate the positioning for the photo? Who knows?

And oh yes, I've already dismissed the idea of it being a tail. Looking at the photograph, it just doesn't fit. Tails, on this planet, seem to be designed from backbone vertebrae and extend out an inch or two above the anus. They have many uses, but most serve at least to stabilize and balance the animal. I suppose there could be some other use on other worlds, but I am fairly certain from it's positioning that it is not a tail. It seems to start coiling and ends at the aliens mid back. At least on a human it would be mid way up ones back. Strangely it seems to end just below the little one's position. Are the two related? The child and the staff? I can not see any reason why they would be and I've gotten no response to my inquiries. Make your guess.

All this having been said, I believe that this alien is a male. Looking closely, you can see what appears to be, (surprise!), male genitalia just below the "wand". This might seem shocking to some but look at the Pioneer plaque! The parts there on the man look pretty much like the parts on this guy and just about the same relative size. Weird huh?

Chapter 34

Aliens, Ancient Gods or Alien Astronauts?

I admit, I ask myself so many questions I sincerely wish I had good answers to. I wonder why I have had so much trouble understanding my alien friends and why they would spend so much time with me attempting to communicate. The time I suppose was just time to get to know the locals. They seemed genuinely interested, in me at least. I'm not sure if they made any other friends in my city. So far I have not heard of any reports of people seeing them. They certainly did not let on to me that they had established an anal exam facility here in town. I'm kidding again. We had zero conversations if you can call them that, about anything to do with physical examination, whether mine or anyone else's for that matter. I suppose it was bonding or something or whatever you wish to call it. The time I spent involved in their activities may have accomplished more than one thing but it certainly brought us closer together. It's far easier to like someone you know something about than to like a perfect alien stranger. Now if they came from our solar system or not I do not know. They are incredibly strange and cool and really don't talk too much. The communication issue is so difficult that I have almost given up on it. I can not understand a spoken language if they have one. I have given myself several really good headaches attempting to do this and do not wish to have another. Imagine a fan with someone standing behind it throwing in small rocks and bits of wood. The sounds which are emitted are not language to you, they are more like constantly changing irritating noise. The mental stuff is nearly as bad. I didn't want it to be bad, and I really tried many, many times to understand what they were doing, and why they were doing it, but I only seemed to get a clear message once in a while. I have said earlier that I also wonder if this problem we had was due to something beyond our control. And that *something* could be what has kept our two races separate for possibly a very long time.

I wonder if these beings are not new to our world, or even my world? Maybe they did arrive post 1940's in metal space ships, which crashed in Roswell, NM. I wonder whether they have been the cause of angst amongst our military and political leaders for just the last few decades or whether they have been interested in us for a much longer period. Possibly they have been here all along. Could it be that this is not my first contact? It could be that they have been here for ages and the lack of ease with which we can communicate has kept our races from becoming friends. Their near invisibility may have kept them hidden. Maybe they are so far advanced that they do not feel the need to speak with us openly or often. Maybe this is Sparky's first contact and the others never shared their past experiences with him? Perhaps they know of some previous contact, but it has been similar to all the many people's journeys to north America before Columbus. The stories would be endless. Everyone says the new land, this Earth is out there. Way, way out there. Way past that group of planets near those two bright stars in the sky. Much further still, it's many light years journey. It's the new lands. And it's rumored that some explorers have been there, but no one has claimed it for their own. Ah, but they can't because they say there are indigenous beings, and they have their own type of primitive intelligence and languages. They are very primitive but sometimes friendly. If you dare to go, watch your step. Travel in a reliable ship. Overall it can be a dangerous place, this planet they call Earth. This world you want to travel to and to explore. Many have attempted to go there, and most of those have never returned from the long journey alive.

Sorry, I just got myself lost in a Viking thing!

The differences in our cultures are vast, and yet we're somehow alike. I kept telling myself that life itself must to be somehow akin to an inter-dimensional virus and also so much more. It cannot be destroyed but only set back. It looks at all opportunities in all dimensions to evolve and advance itself. It tailors its evolution specifically to the environment in which it exists and refuses not to evolve. It may proceed forward either slowly or at supersonic speed, but it will move forward. It reaches out in an attempt to build smarter, more adapted, more intelligent organisms. You cannot burn it, melt it, freeze it, or eradicate it. You cannot pound its essence into dust nor break apart it's pieces. Oh, you can try, and you may stamp out a small portion, but it always survives. It is not just of physical stuff. It is the spark, the animation, the inter-dimensional essence which separates itself from inanimate objects and evolves. In some form, it always survives. It adapts and eventually forms intelligence, and it has it's purpose, and no matter what type of life it may be, the goals are the same. It bridges from one set of realities to another. This is the starting point. We are all from the same family, no matter how different we seem to each other. But all we have to reach out with is our communication skills. Those skills being either verbal or nonverbal, we attempt to bridge the gap and create awareness of our existences, our wants and needs. Yes, communication is tough, but we reach out and attempt contact. We search. We explore. Life has so many options to choose from, and though the basics can be the same, the details can turn out quite differently. Somehow life exists, splits, evolves into separate branches and must somehow eventually communicate between branches to assure the possibility of advancing to it's next evolutionary level.

Take Egyptian hieroglyphs. It must have been mind boggling looking at those beautiful complex images, knowing that there must be a way to unravel their meaning and hear their simple messages. But until the Rosetta stone was discovered there was no way to be sure we were reading them correctly if at all. The Rosetta stone displayed the same text in not only hieroglyphs, but also Greek and Egyptian demotic, bridging the gap between cultures and making translation possible. I suppose, aliens attempting to teach me their language was like Bach attempting to teach a fruit fly to write a symphony. Where's the damn bridge?

Maybe this language, thought process difference thing has been interpreted in the distant past as an offensive gesture, making them seem less friendly and more likely to be the enemies of mankind. If they have been here for a very long time, some groups could actually have been vilified or simply ignored by humans. And with their ability to easily camouflage themselves to an early human population, it might be much easier and safer to remain hidden from view for thousands or even hundreds of thousands of years. Aliens may have thought that the early humans weren't of much interest to them anyway, so if we didn't like them either, they'd just go about their business and try it again when we mature a bit.

But in the meantime........maybe they'll just fly over once in a while to see if we need any inspiration.

So here we have a few aliens which visit our planet, let's say in the distant past. Most can't communicate with us verbally for whatever reason. Maybe a few can. But they all discover they have a knack for nonverbal telepathy and teleimagery. Teleimagery would be communicating mentally through the use of mental imagery. We humans seem better suited to receive.

Looking back through the ages there could to be at least four different small groups out there, but let's leave it at four for now. I'll call them Alien type A, B, C and D for lack of anything better.

Alien Type A

Uses it's telepathy and teleimagery to help groups of humans deemed worthy for whatever reason. The reason may be obvious to us or may be very "alien". High intelligence. Good genetic roots. Maybe we'd been listed as an endangered species on the alien home world. The need to maintain genetic variation within the entire race. Promoting high intelligence and good hygiene? Whatever. Something simple for aliens but possibly something which would make no sense to us. Since we assume the aliens to be more intelligent than us, who knows what they think. But they seem to like us, and when they have nothing better to do or see an opportunity to help, they use their skills to communicate important events. They may also show themselves to us in varied forms, whether human looking or as something else to help get their message across. Let's say they foresee a dam about to break, a huge flood and most humans and animals dying in the process. They inform the humans. Hey Noah, come here. How's your day going? Look, everything is about to get very wet. Trust me, build an ark. The aliens are happy, having done a good deed for the humans. The humans are happy although they don't quite know what to make of the origin of this message. And the aliens go back to doing whatever aliens do when they're not looking after humans. The happy human recipients of the message need a good understandable and believable story of how this information came to them and why anyone should listen to it. The rest of the population as well as them are mystified by earthly events and crave some controlling force which they can speak to about all this weird stuff going on in their lives, so they invent god! How about this? God told me! Great idea!

It covers all the bases. The message is about something humans couldn't possibly know on their own.

The message was delivered by a superior human being type thing in a sort of waking dream or by disembodied voices. Maybe it spoke through a circle of burning grass or a burning bush? Maybe it looked something like a human. The message got across, and like I said, everybody is happy. The aliens have done their good deed, and the humans have survived. No need to shop for new pets. Just possibly these humans will eventually evolve enough and learn enough to figure it out and join us up there.

Up somewhere. Here. Wherever.

A is for **Angels.**

Alien Type B

These guys get credit for most everything. They think it's just too cool to be like gods amongst the human beings. We are just smart enough at this point in our development to see the effects of nature on our planet and our people, and are just dying to give the credit and blame to whoever or whatever beings appear out of nowhere, communicate to us through our minds and disappear when there's trouble. If the sun rises in the morning, they smiled upon us. If the river overflows and crop growth is good this year, they smile upon us. If the oceans well up and destroy our ships, well, they're having a bad day or we've been naughty. They are obviously all powerful and need to be worshiped. Worshiped a lot, and wow, a smart human con man could really get a huge social and economic boost out of being the main connection to these aliens! Yes, the main man. The one who relates (or spins) all the good and bad things that their god has done for (or to) the populace! If the god could reveal himself now and then it would make things a bit more believable, OK?

"Sure," the alien says and goes about eating his beef. I'll even bring out my pet feathered morphing serpent thing for you. Any unknown force of nature or terrific event can be assigned to this god or any other one. I guess you may as well act as two gods at the same time if you can. I mean almost no one can see you anyway and if need be you can speak in a couple of different voices. Just to keep things interesting the aliens throw in a few dream sequences of important upcoming events so that the humans don't get too bored and forget that they're there at this point and that they are much more advanced and smarter than we are.

The humans continue to blame every little thing on their favorite god and the aliens just keep taking the credit, sometimes whether they want it or not.

As news of this successful line of work spreads, more aliens get involved, wishing for a life of luxury and free food. The list of gods grows as the aliens arrive for the gold rush and before you know it you have a whole plethora of stories of gods and happenings and crazy stuff that's gone on which can only be ascribed to, to whoever, gods, angels, jinn and all supernatural beings in general. Of course, none of this is really true, it just works extremely well. Oh they do have some powers we don't, but our imaginations are the root of most of the big physical events. They eat well, make out like bandits, and the sun continues to rise each morning anyway. Everybody's happy!

Eventually, it all begins to break down as human imagination takes over completely, and alien communication does not fill in the gaps properly. The aliens become so complacent that they ignore the obvious signs and continue to feast as if nothing were happening.

So bring me some more deer meat, will you? I love the juicy bloody parts. Now if you humans bring me enough, I'll let you think that the sun comes up in the morning just because I make it so. Hey! Whoa! Hold on. I didn't ask for human blood, just some deer. No, I don't eat human hearts! You're out of deer blood you say? Weeks, month's, years to replenish? What? Don't you guys eat beef? Well then, I'm going north to find some better living conditions in Florida. I really like grouper. You guys can sacrifice all the humans you want. I'm out of here! I want nothing to do with all this killing. Goodbye! The whole thing gets out of hand as the god deserts the humans and consequently they mostly kill each other, all die off or leave the cities.

B is for **Barterers.**

Alien Type C

These aliens are the juveniles and experimenters. The ones who really don't have any evil or misguided intentions, really don't know how to be a god yet and are just running amuck trying to get their heads together. When I was a child, I once had a pet mouse. I was bigger and so much smarter than the mouse. I could do things the mouse could not do. So I did, to the mouse. I picked it up all the time. I spoke to it even though it could not understand. I put it in hot water to give it a bath, and I then put it in the microwave to dry it off. Needless to say, I didn't have a mouse very long. These aliens like to say things to humans to see if they can get them to do something silly or stupid and then laugh when they do. They also just like to see what happens. They like to play pranks, or practical jokes guiding ones thoughts to conclusions which always turn out to be the opposite of what one should do. If they were human kids, you'd probably give-em a time out, but you can't do that with the little alien guys because you can't see them or touch them or influence them. They have a blank check to try to mess with human thought. And once they find a human they like to mess with, they keep going. Usually though they grow out of it or just get bored after a while as most humans won't do what the little prankster aliens tell them to do, unless they want to do it anyway, that is. Sometimes they quit because as they mature, they see the benefits of being kind to other beings as opposed to being bad little aliens. Like I said, most humans won't do the things they inject into their minds, but some weak people may be quite alarmed with the thoughts. Some may claim to hear and see strange images. Some may hear voices of relatives past. Some may hear plants talking to them. Some may think they're crazy. Some may actually be influenced and do some of the idiotic or misguided things suggested by the thoughts in their heads. Some may think they are on drugs. Throughout the ages psychoactive drugs produced by plants and organisms, found commonly in the countryside and woods, made unsuspecting villagers believe everything from the devil and god to evil spirits and witches were speaking to them. But this is not what we are concerned about today. It's aliens. Does the alien explanation account for all human mental illness and other miscellaneous strange thoughts, hallucinations and events? No. Humans have plenty of those all on their own. These aliens really don't mean any harm, but kids just love to play. The ones who just like to experiment, just need to grow up.

C is for **Crappy alien kids.**

Alien Type D

Here's those aliens who don't really get along well with the natives. These alien visitors possibly considered early humans inferior and ugly and really didn't see much point in keeping them around at all. Maybe they see no potential in us because we continue to build social structures which inevitably lead to overpopulation and warfare. We have little or no telepathic ability, communicating like animals most of the time. And speaking of animals we've pretty much destroyed most of the species on this planet. Maybe they think we smell bad and are much larger than them, so we're always getting in the way of their scenic hiking tours. I mean you just make a quick space ship stop to pick up supplies and take on drinking water, and you find some filthy human bathing in the river just upstream of you. The horror! Besides, the humans get freaked out every time we try to think to them, so they must be so ignorant and stupid that they will never make anything of themselves! Most of them would only be good for slaves, and not very good ones at that. This type loathes humans and finds it more fun to terrorize us than ever attempt to work with us or help us. Now telepathy and teleimagery, which I have said seem to be the only way of communicating efficiently for the aliens, has turned out to be just about as much fun as one alien can have for messing about with the stupid humans. Hey! Let's use our telepathy thing to scare the crap out of, irritate and put really nasty, sadistic thoughts and voices in their heads! I see there's a dam about to burst so let's tell them all to let the kids go swimming!

These are the ones who in the Jewish Bible would tell Abraham to take his son Isaac to Mount Moriah and sacrifice him to God! "Type D".

Yes, these are the nasty ones. D for Degenerate, D for Dastardly, D for Destruction.

Now just as Abraham is about the slit his son's throat "Alien A" intercedes and tells him to stop it and go find a nice lamb to bother instead.

I'll just bet Alien "D" gets the crap beat out him by Alien "A" as Abraham is off looking for sheep.

These are the aliens who seriously work against us. The evil ones. They attempt to have us make bad decisions, kill innocent persons and wage wars for little or no apparent reason. They take control of weak minds to wreak havoc, play sadistic games and use every trick in their playbook to torture poor humans and those around them. We might as well call them evil.

Fortunately, they are an extremely small group, a tiny minority of the larger population of aliens. Still, it's mostly all speculation. Maybe, but I base all this upon my real interactions with an alien and it's child. In fact, I have spent almost thirteen years attempting to relate to them and right or wrong have come to these conclusions concerning how this all works in reality. If there is such a thing. Reality, like truth, being simply a matter of perception. I'm not saying that God and the Devil don't exist, but so far I have only seen proof that these guys do, and they could easily play at the same games if you just add in a bit of human imagination and natural coincidence.

D is for **Demons.**

All these alien types are possible because aliens have abilities which naturally give them a huge advantage over the human race. Following the example of the "Invisible Man", aliens can see us and interact with us to some extent, whereas we cannot easily see them and so we also cannot interfere with their actions. And, just like Griffin's anger in the story, The Invisible Man, the aliens temperament and ethical judgment then comes keenly into play. Teleimagery and telepathy are also points which work in their favor. They can speak to some of us in our minds and create or transmit images directly there, whereas we have not yet learned to do this. Communication is pretty much one way and all to their advantage. This could be mostly because we cannot tell if they can hear us or not at any one time, making it more than easy to ignore what we think or say to them. We may think that we can only receive their communications and are not be able to transmit effectively. We may not now understand how to transmit effectively, but we may be able to do so none the less. We may be getting close to being able to do so naturally. If we only knew when they were around us to listen, we might just be able to figure this out. Now, for the first time humans have photographed one. We may possibly learn how to see them whenever they are around. One step at a time.

Most humans have no perceived need or understanding of their technology or communication skills, so having aliens around at all is pointless to them. Most times we cannot even see them and so could care less about attempting to communicate. Furthermore since most humans don't understand alien and might find their communication disagreeable, most aliens may have been seen as a negative influence on, or possibly just background noise to the human race.

If you assume this point to be correct, then you see that aliens could very well have been here throughout history, encountering and interacting with us only to be misunderstood. I'm not saying aliens are the only ones who have these abilities. If you believe all that, there may be spirits, angels, jinn and all sorts of gods and monsters who could do the same things. So there's a mixed up soup of all sorts of things going on, including psychological illness and hallucinogens which could account for the same effects. I think though that there is evidence to conclude that these aliens have been visiting earth since ancient times and have caused little harm to the humans they've encountered. In fact, I would suppose that most really don't seem to care too much about us at all. They may have actually done us some good along the way, possibly by accident, or just by virtue of being what they are, alien and smart. We generally treat them with disdain because they're, well, "alien" for Pete's sake. Sometimes they trick us into believing they look and sound human and then we're accepting of them again. Imagine that. Go figure. Maybe we're ready. Maybe we're not.

Whether you consider them ancient aliens or astronauts, they do seem to be here and have been here. Ancient cultures certainly have seen them. They may have decided to worship them. They have even decided they were so strange and "obviously" powerful that they would kill themselves for them. They probably invented stories and myths about them. Whole religions were probably based upon their appearances. Their interactions with humans may be the stuff of many stories and novels and movies.

Whatever they are, we still do not know precisely, but my bet is that we are now beginning to see that they are a vast resource of unimaginable knowledge. Are we ready for this? Ancient peoples obviously were not, but were still worthy of some small form of acceptance and direction. We are the human race. We will survive. Survival is in our nature. I believe they understand this and so have been here for ages upon ages with us, watching and waiting for the right moment. Just one precise point in time. The correct time. The correct time to make real contact.

Chapter 35

Aliens and Coincidence

Example of coincidence: I went to a party last night and thirteen of the men there were named Andy.

Aliens love coincidence. Multi-levels of complexity. Plethoras of interlacing gobble-t-gook. Eleven dimensional chess. Humans usually don't see things that way, or at least didn't for thousands of years and saw coincidence as a sign of the work of their gods. Then we got smarter. Gods faded, science and mathematics expanded, and our minds began to fill more with probabilities than coincidence or godlike interventions. But still there is that science so advanced and strange as to make us believe in that magic thing again.

Why is the message contained in a photograph? Probably because some of our earliest transmissions into space contained visual information. I Love Lucy. The Olympics in Germany. Cartoons. The Honeymooners. Why didn't we think to send out something a bit more intellectual? Who knows, and if aliens picked up these signals and have watched the weekly programming, they probably think we are a pretty funny race. We transmit daily news, kids cartoons, horror movies, romance, comedy, mysteries and science programs all at once. Maybe they see this as an attempt at multi-levels of complexity and coincidence. I sort of hope we gave them a migraine or two! Poor aliens. If we had transmitted an encyclopedia of earth, out into space, they might have responded with the same, but no, we transmitted The Three Stooges. After a few decades of TV, they obviously realize we are now progressing in our obsession with the ridiculous at a very fast pace. Hopefully they are mature enough to see through it all.

Are the aliens, which have possibly been here for thousands of years, here by choice? Mostly yes, I think. Seeing as how some were possibly left here, crashed here, or were abandoned here many thousands of years ago and have come to call this place, this earth their home, means that we could be their adopted families! My photograph stands on its own. The negatives are in my possession. There is no difference between the two. Sparky stands proudly beside my dog displaying his child, his tools, his works of art and obviously asking that I be his friend. His hands outstretched holding gifts are ready to use their contents to please and amaze me. I am still in awe of this initial interaction, and I suppose the feeling will never go away. I responded to his gifts with an offering of my own. I make myself available for their tests. I open my home to them, and we have become friends. Time has proven me correct that they have meant no harm, no intentional evil deception, and have only good intentions But there's still more.

Chapter 36

Exchanging Gifts and Ideas

What would you have to give to an alien race, a human race, if you encountered them? Would you bear gifts of beads and trinkets? Would you hand them a stack of intergalactic cash? How about some tasty alien snacks? Probably not, but you would have something to offer, and if your race has been visiting their planet for hundreds or thousands of years, you would take something with you that might relate this to them. It would tell them that you have been present and have done them no harm. Solid objects would likely be out of the question since you are presenting yourself in a sort of semi-invisible holographic-like state. And you certainly don't want to share something which might disclose advanced technology to them, which they might use to blow themselves up with. These humans seem extremely smart, and the smallest hint might be too much. But, you want it to be known that this is a special gift you are offering to them, and it is not just some small bottle of shampoo or something else you happen to have in your travel bag. You must make the correct choice in gifts. You have one chance at this. You must be able to hold it in your hands. A gift of knowledge perhaps? Yes!

But the information should not be of past human events like wars or earthshaking destruction upon this planet. It should be something which speaks to your partnership with and peaceful observance of their race. A map perhaps? But of what? The solar system? At this point, you are pretty sure that is not going to surprise or impress them. They are already well aware of that. All known black holes in the universe? Too complex. It must be something simple, unique and special. Something along the lines of the message on the voyager plaque? Yes! And you must be able to easily hold it in your hands. Your outstretched hands. For the photo. So what sort of simple map would you choose? A map to your home might not be wise considering your experience with *this* race so far. A map of time perhaps? With them? Yes! How about a simple pictorial map of their past ancient history?

In Sparky's hands, you will see a stylized modern human face, behind it sits an Egyptian face, a princess, a pharaoh's wife perhaps, followed by a giant, an enlarged (ancient) human skull. Look deeply. They're there.

I believe this to be just what it seems, a record of their race's interaction with ours and the evolving forms of the human face. What a wonderful gift! It tells us that they have been here all along, and have encountered humans at various points in history and have existed with us in peace.

Another puzzle piece in place. Namaste.

Human face with spiky head.

Egyptian face.

Giant head.

Oh, and as for the singing in your sleep, sorry, but you'll just have to wait for the next book to find out about that one. I may also tell you my secret to blocking out annoying alien telepathy as well. Maybe.

I would hang on to this book if I were you. There may be more important images uncovered. We are all in this together.

Chapter 37

Contact

Some aliens have been here helping the native peoples of North America for a long time, possibly so long that it can not be remembered, perhaps teaching them to hide within the earth when the lands became frozen and barren. Legends. I believe that what we are seeing here is an ancient race, one that has been here for thousands of years, whether technological or natural in their ability to confound our senses and manipulate our perceived reality. They are here now and consider the earth their home as much as the human race does. I sense that they are not our gods and are not deities. They do not want us to ever portray them in that light again. The transgressions of the past. The Aztecs. The blood lust, the death, the sacrifices, the worship, should never be repeated, ever again. If that happens, they will abandon our culture and disappear once more. I feel it. I sense it. I seem to visualize it.

It has occurred to me that it's easy to get lost in too much speculation. Unfortunately, there are only my own experiences and the photograph on which to base many of my theories. Communication was quite sketchy to say the least, and did not help to clarify many assumptions about them at all. There is possibly some additional supporting evidence which I will only mention at this time because I have no idea how much importance I should give it or how reliable it might prove. The photo I have verified is absolutely untouched, unaltered and totally authentic except for text on the cover.

I do have a couple of items which are yet to be researched. One is a small flat piece of metal-like substance which I discovered afterwards in the dirt where the photo was taken. It looks like a piece of molten putty which has coalesced into a metal puddle! It is non-magnetic and has a strange clunk to it when tapped on another surface. I have searched the area repeatedly and found lots of seeds and dried up vegetation. I have dug into the ground repeatedly and have found various items, but not one was similar to the glob of goo-metal, and it was laying on the surface of the ground, not buried beneath. It looks like it had been there for some time, possibly several years or more. I found it in 2010 some years after my first visual experience with the large orb. Coincidence again? Why are there no other globs of metal there, or anywhere else nearby? I will someday be able to take this item to a lab and hopefully determine it's composition and origin, but from what I have heard, there have been like items found near UFO sites around the world. Hopefully I will be able to give you more on this at a later date. I also have a photograph of a light source which might be related to the orb which appeared on nights near the date of the alien photo. I do not put much stock in photos of points of light which appear in the night sky, and I must say that I also have yet to do any research on this photo as well.

The last piece of potential evidence is another very strange group of photos. They were taken near the time that the alien photograph was taken and also when the bright light was photographed, but this group of photos is of a clock. The clock itself is sitting still. Each successive photo in order shows the clock hands moving forward a few seconds at a time. I have looked again and again at the series and many times almost thrown them away, not certain why I would have taken them. I barely even remember taking these photographs. The light is strange but not gooey. There appears to be some hand shake in the shots. The clock was a cheap model I threw into the trash a few years ago. It had stopped working and never started again. I still have the negatives which appear to be exactly the same. Same lighting and some blurs. But there's a lot of them. Almost a whole roll.

I've said before I am no photographer and don't understand the simplest of concepts, or at least I didn't at the time. But a suspicion slowly crept it's way into my head. What if the sequence numbers on the roll of film was not the correct order in which I took the photos? From what I had understood the photos should have been printed and numbered in the order they were taken. I examined the photos to see if I could ascertain anything which would tell me what order they were in. The correct order. The photos show the hands on the clock moving steadily forward. But something is off. Could the sequence be backwards? Was the photograph numbered one, actually the last photo I took. If that were the case, I may have taken a series of photographs of time going in reverse. But, time moving backwards! Einstein would have dropped his pipe! "Impossible!" I said. This can not be happening, and something else is off. I mean it must be wrong, and the sequence is correct because time always moves forward. Doesn't it?

Well, I can't prove it just yet but, I believe the first photo I took was actually the last to be printed, and numbered last. Possibly because of the way the film wound around the spindle. It may make some sense. I think I remember that on that night, a night with strange lights, I somehow lost my way. I don't know how to describe it. Like being lost in your own front yard. But while struggling to regain my composure I had the mental focus to look at a clock and notice that it was behaving strangely. It was to my amazement ticking backwards! Why? Not sure. But, I think I can prove that the order was as I have said showing the clock was indeed ticking backwards! This explains why I would waste a roll of film taking photos of a clock just sitting there doing it's own time thing. Do I remember what I was doing that night? Not exactly. Only bits and pieces remain. I have lost most of that memory, but I have recalled that at some time that night the aliens may have showed me the inside of their "orb". How this all fits together I have no idea. I also have not yet had this researched, but I can attest to the fact that it is true as far as I can tell. So I will leave it at this.

A friend asked me if I would ever submit to hypnotherapy to attempt to recall the entire experience, but I have a natural distrust of experts conducting hypnotherapy or utilizing tests to uncover facts or prove an event one way or another. All the tests in the world won't make a difference if the alien's technology appears to us as magic. I do not wish to possibly overwrite my own true experiences with imaginations or doubts. All these tests and techniques have been proven at times to be unreliable and troublesome so I will not cloud my memory with these, whether physical or psychological invasions of my mind. You can't put it back together once it's been perverted. Sparky has also advised against all such actions. I can only tell you what I have experienced and how the aliens treated me. I am very pleased to be the one they chose for this contact. And after almost thirteen years of living with their presence, I am well and I am happy.

There is a definite purpose in this event. It's not just random, and it's not accidental. It's not pareidolia, and it is not confabulation. It is neither self delusion nor schizophrenia. Sparky is purposefully presenting a full and correct depiction of their race just as we did on our probes sent deep into space so many years ago. Coincidence? Perhaps, but we should not overlook the possibility that this is the response we have been hoping for. They are out there in space, and they are here now as well. As a race, they wish to contact us and to exist with us. They are speaking to us and telling us now that we are not alone.

As two intelligent races, can we handle this interaction? Can we get past the tendencies towards hostage taking, technology theft, and genetic material collection? What will it take to put one foot in front of the other to bring both parties together in peace? I'm not certain it will ever happen, but here it is, the first footfall by another intelligent culture. One being has taken a step and proclaimed I am not afraid! We are also here! We are like you in so many ways, and we too are friendly! He stands in the open now, naked and nearly defenseless. He waits.

My best guess is that they arrived in late 1999. They appeared to me in the year 2000 and stayed through 2012. Where they are now, I cannot say. They have gone. In the early spring of 2013, they left. Sparky thanked me for my patience and said that he is concerned, but pleased with us. He also said that he would return in time to check again. He related that we will have troubles, but if just one human can show respect and friendship, can overcome fear and intimidation, without retaliation or hatred, and demonstrate that the human race is evolving mentally towards... (no translation), then more will follow, and so he will leave us for now in peace to find our way through the bitter maze of challenges ahead of us. He said that this is now our time, our revelation, a time of terrible trials and wonderful new beginnings we have yet to fully embrace or understand, but will look back upon clearly and proudly.

At times, everything just seems to swirl around.

I sometimes feel that my experiences have been similar to a dance, a dance to a beautiful melody which sails and floats on the wind, a dance of ages, a dance with friends. Possibly I have been dancing with something more powerful than can be imagined, and unknown, possibly in my own mind, only in my own thoughts, but perhaps, just perhaps I have been dancing with an alien spirit.

And that is a story for another day.

Completed on Thanksgiving Day, 2013 in remembrance and in honor of my guests, and Greggor who recently passed away.

BIBLIOGRAPHY

Allen, James P. Middle Egyptian: An Introduction to the Language and Culture of Hieroglyphs. Second Edition. New York: Cambridge University Press, 2010.

Anderhub, Werner and Hans Roth. Crop Circles: Exploring the Designs & Mysteries. New York: Lark Books, 2000.

Anderson, Roland C. Octopus: The Ocean's Intelligent Invertebrate. Portland: Timber Press, 2010.

Banister, Keith. The Encyclopedia of Aquatic Life. Oxford: Equinox Ltd., 1985.

Barrett, Deirdre. The Encyclopedia of Sleep and Dreams: The Evolution, Function, Nature and Mysteries of Slumber. Santa Barbara: Greenwood, 2012.

Beech, Martin. The Physics of Invisibility: A Story of Light and Deception. New York: Springer Science + Business Media, LLC., 2012.

Bernstein, Peter L. The Power of Gold: The History of an Obsession. New York: John Wiley & Sons, 2000.

Birnes, William J. The UFO Magazine: UFO Encyclopedia. New York: Pocket Books, 2004.

Edgerton, Harold. Stopping Time: The Photographs of Harold Edgerton. New York: Harry N. Abrams, Inc., 1987.

Einstein, Albert. The Meaning of Relativity. London: Routledge, 2003.

Forbes, Tom. The Invisible Advantage Workbook. Ghillie Suit Construction made Simple. Colorado. Paladin Press, 2002.

Friend, Tim. Animal Talk: Breaking the Codes of Animal Language. New York: Free Press, 2004.

Greene, Brian. The Elegant Universe: Superstrings, Hidden Dimensions, and the Quest for the Ultimate Theory. New York: W.W. Norton & Company, Inc., 1999.

Hawking, Stephen. A Brief History of Time. New York: Bantam, 1998.

Howe, Linda Moulton. An Alien Harvest. Linda Moulton Howe, 1989.

Jay, Joshua. Magic: The Complete Course. New York: Workman Publishing, 2008.

Kaku, Michio. Parallel Worlds: A Journey Through Creation, Higher Dimentions, and the Future of the Cosmos. New York: Doubleday, 2005.

Kaku, Michio. Physics of the Impossible: A Scientific Exploration into the World of Phasers, Force Fields, Teleportation, and Time Travel. New York: Doubleday, 2008.

La Berge, Stephen. Lucid Dreaming: A Concise Guide to Awakening in your Dreams and in Your Life. Colorado: Sounds True, Inc., 2009.

Lassieur, Allison. Albert Einstein: Genius of the Twentieth Century. New York: F. Watts, 2005.

Lorentz, Hendrik Antoon: The Einstein Theory of Relativity: An Explanation and Appreciation. WLC, 2009.

Lykken, David. A Tremor in the Blood: Uses and Abuses of the Lie Detector. New York: Plenum Trade, 1998

National Geographic Essential Visual History of World Mythology. Washington, D.C: National Geographic Society, 2008.

Newark, Tim. Camouflage. New York: Thames and Hudson, 2007.

Peterson, James. Fish and Shellfish: The Cooks Indispensable Companion. New York: William Morrow Cookbooks, 1996.

Rock, Andrea. The Mind at Night: The New Science of How and Why We Dream. New York: Basic Books, 2004.

Rossing, Thomas D. Light Science: Physics and the Visual Arts. New York: Springer-Verlag New York, Inc., 1999.

Sagan, Carl. Drake, F D. Druyan, Ann. Ferris, Timothy. Lomberg, Jon. Sagan, Linda Salzman. Murmurs of Earth: The Voyager Intersteller Record. New York: Random House, 1978.

Stanley, Sadie. The New Groove: Dictionary of Music and Musicians. Second Edition. New York: Macmillan Publishers Ltd., 2001.

Wells, H.G. The Invisible Man. New York: Signet Classic, 2002.

Westrup, J. A. The New College Encyclopedia of Music. New York: Wm Collins Sons & Co Ltd., 1959 & 1976.

White, Laurie. Infrared Photography Handbook. Buffalo, NY: Amherst Media, Inc.,1995.

Williams, Raymond. Contact: Human Communication and its history. New York: Thames and Hudson, 1981.

Recommended Reading:

Adams, Douglas. The Ultimate Hitchhikers Guide to the Galaxy. New York: Ballantine Books, 2002.

.....and if you ever see visitors from far away lands..... show them hospitality.

www.ingramcontent.com/pod-product-compliance
Lightning Source LLC
Chambersburg PA
CBHW051451170526
45166CB00001B/205